THE HUMAN BODY

THE
HUMAN BODY

AN ILLUSTRATED GUIDE TO ITS FORM, FUNCTION AND CAPABILITIES

ROBERT SNEDDEN

SIRIUS

SIRIUS

This edition published in 2024 by Sirius Publishing, a division of
Arcturus Publishing Limited,
26/27 Bickels Yard, 151–153 Bermondsey Street,
London SE1 3HA

ISBN: 978-1-3988-4333-2
AD010138UK

Printed in China

Contents

Introduction

From the earliest times people have doubtless wondered about how exactly their bodies functioned. The effects of injury on the human body might have laid bare what was going on 'under the skin' but there was little understanding of how the unfortunately revealed inner organs might work.

Crude anatomical drawings dating back at least 25,000 years have been found in caves in Europe, Asia, Africa and Australia. The ancient Egyptian practice of mummification, dating back more than 3,500 years, meant there was some understanding of the human body's inner workings, though mummification was carried out for religious purposes, not with the aim of acquiring anatomical knowledge. There is some evidence that dissections were carried out as early as the 6th century BC in India, though this would have been done without the dissector's hand touching the body directly as contact with a dead human body was forbidden by religious laws. For most of recorded history the dissection of

human bodies has been forbidden by law and by religious belief and often punished harshly.

The word anatomy, from the Greek *ana-tome*, meaning 'cutting up' or 'dissection', was first used in the 4th century BC by Aristotle, who dissected animals. Knowledge of anatomy was presumed to be important for the practice of medicine, but this knowledge was gained from animal experimentation or by studying wounds and injury to humans. There is no evidence that the physicians of ancient Greece practised the dissection of human cadavers at this time.

It is possible that the Egyptian practice of mummification might have had some bearing on the acceptance of working with cadavers in 4th century BC Alexandria. Herophilus, who is sometimes described as the 'father of anatomy', is the first person known to have performed systematic dissections of the human body. He is said to have carried out at least 600 human dissections and was rumoured to have practised vivisection on convicted criminals, though there are doubts as to the authenticity of this claim. Dissection undoubtedly gave Herophilus insights into the workings of the human body that had been

A stone relief at the entrance to the Nouvelle Faculté de Medicine in Paris commemorates the first known dissections of the human body carried out by the Greek physician Herophilus in 4th century BC Alexandria.

denied to previous anatomists. His work on the nervous system was particularly accomplished. He was the first to distinguish between motor and sensory nerves and also between spinal and cranial nerves and described structures in the brain, such as the cerebrum, cerebellum and meninges.

Herophilus was working at a time when dissection of the human body was generally looked on as an act of desecration. After his death, a new school of medical thought came to prominence, which maintained that a dead human body was so dissimilar to a living one as

The process of mummification, which involved the removal of internal organs, gave the Egyptians some inside knowledge, as it were, of human anatomy, though not of how the organs functioned. They believed, for instance, that it was the brain that pumped blood around the body rather than the heart, which was responsible for thought and emotion.

Above: Galen, depicted here with the serpent-entwined Rod of Asclepius, a Greek deity of healing and medicine, was one of the most prominent surgeon-physicians of the Roman world. His understanding of anatomy dominated Western medical science for over a thousand years.

to render it of little use in medicine. The practice of dissection was abandoned for the next 1,800 years, restarting only in the mid-16th century.

Another of the great anatomists of classical times was the Roman physician, Galen. His works on anatomy were consulted by physicians for the next 1,300 years until they were challenged by the discoveries of the 16th century. Galen gathered much of his knowledge by performing vivisections on animals; his studies of pigs and apes providing the basis for much of his beliefs regarding the inner workings of the human body. As chief physician to the gladiator school at Pergamon, Galen had the opportunity to study wounds of all kinds and severity and get glimpses of the hidden secrets of the body without himself wielding the knife.

Medicine, and scientific progress generally, made little progress for over a thousand years in the West until the period of history that came to be called the Renaissance saw an easing of the legal and cultural restrictions on dissecting cadavers, and knowledge of human anatomy began to move forward again.

One of the prime movers in this resurgence was the great polymath Leonardo da Vinci (1452–1519) who began his exploration of the 'marvellous human machine' in 1506 with his dissection of a 100-year-old man. In all he dissected around 30 corpses and, in the winter of 1510–11, he compiled a series of more than 240 exquisite drawings and over 13,000 words of notes setting out his findings.

Opposite: Leonardo da Vinci's skill as an artist and meticulous attention to detail enabled him to produce some of the finest anatomical studies of the human body ever seen.

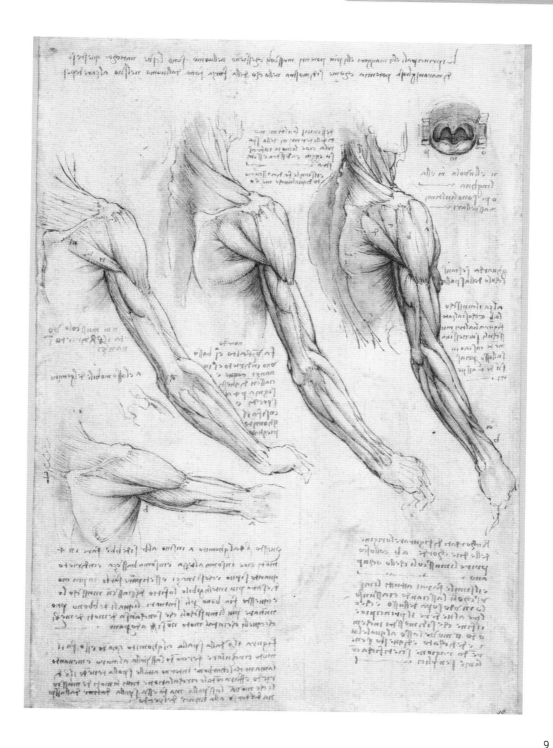

Although Leonardo's interest in anatomy originated in a desire to improve his art, it held a fascination in its own right. His dissections and drawings of muscles, nerves and blood vessels were painstaking. In particular he investigated the structures of the brain, heart and lungs; described conditions such as atherosclerosis and cirrhosis of the liver for the first time; and made the deduction that it was the retina that was the light-sensitive part of the eye, and not the lens as had been previously believed. He studied the coronary vessels and valves of the heart, making a glass model of the aorta to study the flow of blood through the aortic valve.

Leonardo's notes and drawings went unpublished and undiscovered for centuries. It is unfortunate that he never considered himself to be a professional anatomist and didn't share his findings. Today his anatomical drawings are considered to be one of the finest achievements of Renaissance science and a benchmark in modern scientific illustration.

One man with perhaps greater claim than Herophilus to the title of 'father of anatomy' was Andreas Vesalius. Born in Brussels in 1514, Vesalius was one of a long line of physicians and studied at the medical university in Paris at a time when fresh interest was being shown in the science of anatomy and the accuracy of the classic works was being questioned.

Vesalius was determined to carry out his own anatomical investigations. He became Professor of Surgery at the University of Padua where he taught anatomy. He transformed the discipline, insisting on the importance of actual dissection, by both teacher and pupil, and used his experience to reveal numerous inconsistencies in the teachings of Galen. Galen's writings, which had previously dominated the study of the human body, had been based primarily on animal dissections, which meant there were inevitable errors in his suppositions. Vesalius' challenges to Galen were bitterly attacked by traditionalists; yet he continued to champion the evidence of his experience. In 1542, he published his anatomical masterpiece, perhaps the most famous text in medical history, *De humani corporis fabrica, libri*

Andreas Vesalius challenged the entrenched medical beliefs that had held sway since Galen's time, insisting that observation and experiment were more important than accepted dogma.

Opposite: The title page of Vesalius' De humani corporis fabrica illustrates the often theatrical nature of a dissection. The first anatomy theatre opened in Padua in 1595.

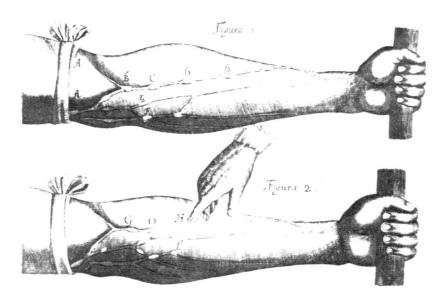

This illustration from William Harvey's De Motu Cordis *shows a demonstration that blood in the veins of the arm flow towards the heart but not away from it.*

septum (Seven books on the fabric of the human body), a study of the human body that ultimately set anatomy, and medicine, on to a new course of observation and experiment. Each of the seven volumes covered a particular topic: skeleton, muscles, vascular system, nerves, gastrointestinal system, heart and lungs, and the brain. It quickly became the standard reference work for anatomy throughout Europe.

One of Galen's mistaken assumptions was that blood was made in the liver before being carried by the veins to all the organs of the body, which attracted the blood to themselves. According to Galen, the blood did not circulate but was consumed by the body, which meant that it needed to be constantly replenished. There are suggestions that physicians of other cultures had grasped the nature of the circulation of the blood. A Chinese medical manual, written 2,600 years ago, stated that 'all of the blood in the body is pumped by the heart, completes a circle and never stops moving.' Arab doctor Ibn al-Nafiz, writing

in the 13th century, described what he called the 'small circulation' of the blood from the heart to the lungs and back, but he was largely ignored.

William Harvey, born in Folkestone, Kent, in 1578, studied at the University of Padua under Fabricius, the first person to clearly describe the valves in the veins and a major influence on Harvey's thinking. Harvey studied the works of Vesalius and learned to adopt an experimental approach to learning about the human body by dissection. Harvey carried out experiments and observations, most of which was accomplished by vivisection, on the cardiovascular systems of a variety of animals, ranging from worms to birds and dogs. As a teacher of anatomy, he also had ample opportunities for carrying out autopsies and dissections of human cadavers. Through a series of painstaking investigations – watching the blood flow from sectioned arteries and veins; watching the beat of a living heart and the direction of blood flow and the action of the cardiac valves – he slowly built up a body of evidence.

In particular, Harvey was impressed by the sheer volume of blood present in the body. There was no way that it could be constantly replenished by the liver. It had to be the case that the same blood was being used over and over again, passing through the body from the arterial to the venous system and the heart was the pump that kept it all moving. He surmised that the blood that had flowed through the arteries returned to the heart via the veins, though he lacked a microscope that would have revealed the capillaries that joined the two.

The invention of the microscope revealed new details of anatomy. In 1661, Italian biologist Marcello Malpighi succeeded in locating the capillaries that had eluded Harvey. Around the same time Danish physician Thomas Bartolin discovered the lymphatic system.

In the 20th century new techniques and technology opened up the exploration of the living human body, transforming anatomical studies and allowing physicians to scrutinize the body's workings without having to cut it open.

In December 1895, Wilhelm Röntgen

Harvey was court physician to both King James I and King Charles I. Here, in a painting by the artist Ernest Board, he is depicted demonstrating his theory of the circulation of the blood to Charles I.

The invention of the microscope at the beginning of the 17th century gave anatomists the ability to begin exploring the workings of the human body in never before seen detail.

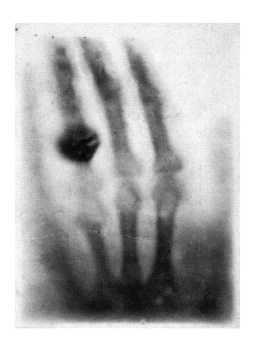

The first X-ray image, taken in 1895, shows the wedding ring and ghostly skeletal hand of Wilhelm Röntgen's wife Anna's hand. Reportedly, she exclaimed 'I have seen my death!' and refused to set foot in the laboratory again.

discovered a new type of ray that could pass through most substances, including through the soft tissue of humans, but not bones and metal objects. The rays would form an image on a photographic plate and one of the first images Röntgen captured was of the bones in the hand of his wife, Bertha. Röntgen's discovery of what came to be called X-rays caught the attention of both science and the general public around the world. Within weeks of the discovery being announced medical radiographs were being made to guide the work of surgeons in Europe and the United States. Just a few months after the world became aware of their existence, X-rays

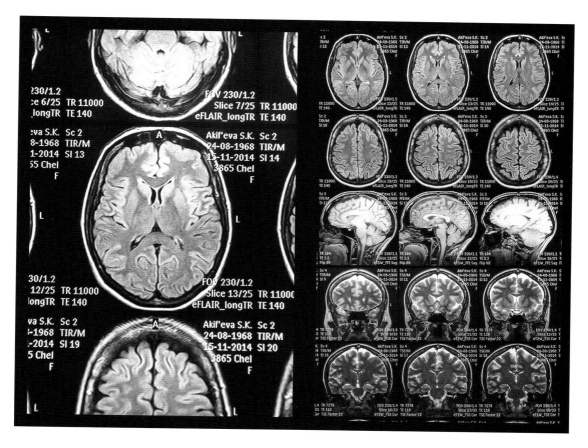

Modern scanning techniques, such as magnetic resonance imaging, allow physicians to observe the inner workings of the human body without resort to invasive surgery.

were being used by battlefield physicians to locate bullets in wounded soldiers without the need for lengthy and dangerous exploratory surgery. In honour of his discovery, Röntgen would be awarded the first Nobel Prize for Physics in 1901.

The pioneers of X-ray radiography little suspected the dangers involved in this ground-breaking discovery – radiation burns were common to operators and patients alike. Today, patients are minimally exposed to low level radiation, making X-ray scanning virtually risk-free.

In the mid-1970s hospitals began to introduce computed tomography scanning, which uses X-rays to produce 3D images of the inside of the body along with magnetic resonance imaging (MRI) scanners, which use strong magnetic fields to produce detailed anatomical images. Unlike CT scanners, MRI scanners do not generate radiation and are essentially harmless to the patient. These new technologies, along with inventions such as ultrasound scanning and the electron microscope, mean that we now understand how the human body works in unprecedented detail.

From Cells to Systems

From Cells to Systems

The structural and functional unit of all living things is the cell. Some organisms, such as bacteria and protozoa are single-celled, or unicellular, others, such as humans, are multicellular organisms formed from conglomerations of trillions of cells of different types working together.

INSIDE THE CELL

Just as different cell types perform unique and essential functions to ensure the wellbeing of the organism as a whole, so structures within each cell, called organelles, or 'little organs', work together to keep each cell healthy and functioning.

The boundary between the interior of the cell and the surrounding environment is the plasma membrane, a double layer of phospholipids in which proteins are embedded. The plasma membrane acts like a gatekeeper, controlling the passage of molecules into and out of the cell.

Far from being a simple 'blob of life' a cell is a complex structure housing, in most cases, a central nucleus where the genetic material is stored, surrounded by a membrane-encased cytoplasm housing a variety of organelles that perform tasks such as supplying energy (mitochondria) and assembling proteins (ribosomes).

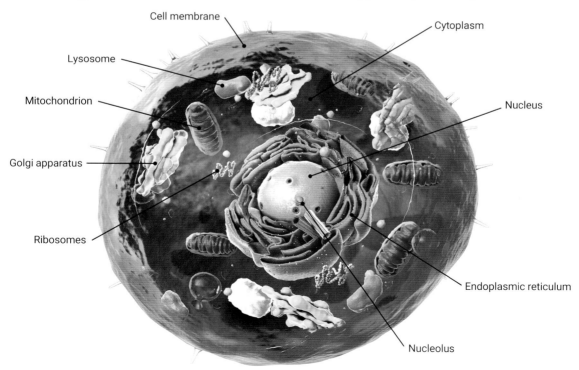

Cell membrane

Lysosome

Mitochondrion

Golgi apparatus

Ribosomes

Cytoplasm

Nucleus

Endoplasmic reticulum

Nucleolus

The plasma membranes of cells that specialize in absorption, such as those lining the small intestine, have finger-like folds or projections called microvilli which increase their surface area.

Within the plasma membrane is the cytoplasm, made up of the gel-like cytosol in which the cell's organelles are suspended. The cytoplasm is mostly water with chemicals such as glucose and other sugars, amino acids, nucleic acids, fatty acids, and ions of sodium, potassium, calcium, and many other elements dissolved within it.

The most obvious of the cell's organelles is the nucleus. Structures within the nucleus called chromosomes are made up of DNA, the cell's hereditary material, which is basically a set of instructions for assembling proteins. Every species has a specific number of chromosomes; in humans the chromosome number is 46. Chromosomes are only visible and distinguishable from one another during cell division.

DNA in the nucleus is transcribed into messenger RNA (mRNA), which leaves through the nuclear membrane and travels to the cell's ribosomes. Ribosomes are the cell structures responsible for synthesizing proteins. They appear either as clusters or singly, floating in the cytoplasm. The ribosomes translate the instructions encoded in the mRNA into the specific order of amino acids, the building blocks of proteins, that make up a particular protein. Protein synthesis is one of the essential functions of cells; hormones, antibodies, enzymes and various structural components of cells are all formed from proteins. Cells such as those in the pancreas, where several digestive enzymes are produced, are particularly rich in ribosomes.

The main energy-carrying molecule of the cell is adenosine triphosphate (ATP) which is produced in the oval-shaped mitochondria. ATP is made using the chemical energy stored in glucose and other nutrients through the process of cellular respiration. This requires oxygen and produces carbon dioxide as a waste product, which is expelled when we breathe out. High concentrations of mitochondria are found in muscle cells.

TISSUE TYPES

The human body is composed of around 200 distinct types of cell, varying enormously in their structure and function. Some of the cells of the human body are not attached to other cells and act individually. Red blood cells are a good example of this. Other cells are organized into tissues. They are attached to, and act in concert with, other similar or related cell types to perform specific functions within the body.

There are four broad categories of tissue in the human body: epithelial, connective, muscle and nervous tissues. Epithelial tissue, or epithelium, makes up the sheets of cells lining internal cavities and covering exterior surfaces of the body. Some glands are also formed from epithelial tissue. Connective tissue protects, supports and binds together the cells and organs of the body. There are three major types of muscle tissue: skeletal (or voluntary) muscle, smooth muscle and cardiac muscle. Muscle tissue contracts to produce movement. Nervous tissue produces nerve impulses in the form of electrochemical signals that facilitate communication between different parts of the body.

Epithelial tissue

Epithelial tissues are classified according to the shape of the cells that make them up and the number of layers of those cells.

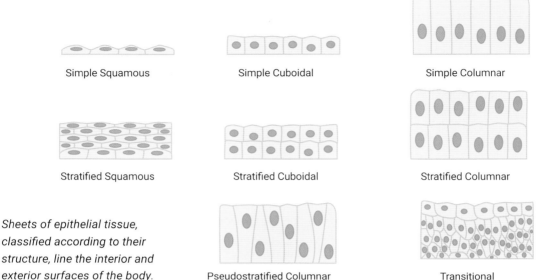

Simple Squamous   Simple Cuboidal   Simple Columnar

Stratified Squamous   Stratified Cuboidal   Stratified Columnar

Sheets of epithelial tissue, classified according to their structure, line the interior and exterior surfaces of the body.

Pseudostratified Columnar   Transitional

Simple squamous epithelium looks like thin scales in appearance and is found lining capillaries and the small air sacs of the lungs where rapid passage of chemical compounds takes place. It also makes up the mesothelium which secretes the serous fluid that lubricates the internal body cavities.

Simple cuboidal epithelium is involved in the secretion and absorption of molecules requiring active transport and is found in the lining of the kidney tubules and in the ducts of glands.

Simple columnar epithelium makes up most of the digestive tract and some parts of the female reproductive system. It is also involved in the absorption and secretion of molecules using active transport.

Ciliated columnar epithelium is composed of simple columnar epithelial cells which also have cilia on their apical surfaces. These cells are found in the lining of the fallopian tubes, where they assist in the passage of the egg, and in parts of the respiratory system, where the cilia help to remove particles.

Pseudostratified columnar epithelium appears to be stratified, or layered, but in fact consists of a single layer of irregularly shaped and differently sized columnar cells. These cells are found in the respiratory tract and some may have cilia.

A stratified epithelium consists of multiple stacked layers of cells. It protects against physical and chemical damage.

Stratified squamous epithelium is the most common type. The uppermost cells appear squamous, or scale-like, whereas the basal layer is formed from columnar or cuboidal cells. The top layer of the epithelium may be covered with dead cells containing keratin. Skin is an example of a keratinized, stratified squamous epithelium. The lining of the oral cavity is an example of a stratified squamous epithelium where the upper layer is not keratinized.

Transitional epithelium is a type of stratified epithelium that is found only in the ureters and bladder of the urinary system. When the bladder is empty, the epithelium has cuboidal apical cells with umbrella-shaped surfaces. As the bladder fills, the epithelial cells change in appearance from cuboidal to squamous. The cells are stretched out and less stratified when the bladder is full and distended and thicker and multi-layered when it is empty.

Glandular epithelium, as the name suggests, form glands, structures that synthesize and secrete chemical substances.

Connective Tissue

The primary function of connective tissue is to connect and support the other tissues in the body. Unlike epithelial tissue, which is composed of closely packed cells with little or no space in between, the cells in connective tissue are held in a matrix of extracellular material. The major component of this matrix, which is produced by the cells embedded in it, is a ground substance, which can be fluid and gel-like, but can also be mineralized and solid, as it is in bones. The matrix is often crisscrossed by protein fibres and plays a major role in the functioning of the connective tissue.

The three elements of cells, ground substance and protein fibres are typical of connective tissue, which can be divided into three broad categories according to the characteristics of their ground substance and the types of fibres embedded within the matrix. Loose connective tissue and dense connective tissue, which together make up

Connective tissue, as the name suggests, connects and supports other tissue types in the body. The cells are held in a matrix of material which can be liquid, like blood, or a dense solid, like bone.

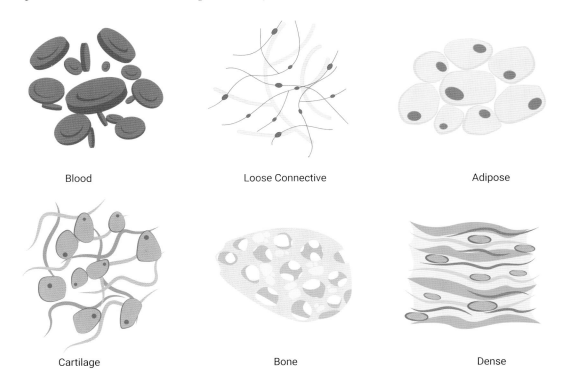

Blood

Loose Connective

Adipose

Cartilage

Bone

Dense

connective tissue proper, both have a variety of cell types and protein fibres suspended in a viscous ground substance. In loose connective tissue, the fibres are loosely organized with large spaces in between, whereas the reinforcing bundles of fibres in dense connective tissue provide tensile strength and elasticity. Supportive connective tissue is characterized by densely packed fibres and a few distinct cell types within a matrix. In the form of bone and cartilage it provides structure and strength to the body and gives protection to soft tissues. Fluid connective tissue, in the form of lymph and blood, consists of a variety of specialized cells circulating in a watery fluid of dissolved salts, nutrients and proteins.

The contraction of muscle cells provides the pulling force that makes the parts of the body move.
The smooth muscle that lines organs such as the digestive tract is not under conscious control, nor is the cardiac muscle that ensures the steady beating of the heart. Only skeletal muscle obeys conscious commands from the brain, allowing us to perform a variety of physical activities.

Muscle Tissue

Muscle tissue is characterized by cells that respond to a stimulus by contracting and generating a pulling force causing movement in the body. Some muscle movement is voluntary, meaning it is under conscious control. For example, when you decide to go for a walk. Other movements are involuntary and not under conscious control, such as the reflex action of quickly removing your hand from a hot surface.

There are three types of muscle tissue classified according to structure and function: skeletal, cardiac and smooth.

Skeletal, or striated, muscle is attached to bones and its contraction makes possible voluntary movements of the body such as locomotion and facial expressions. Forty per cent of the average adult's body mass is made up of skeletal muscle. As a by-product of their contraction, skeletal muscles generate heat and so also play an important role in maintaining the body at an optimum temperature. Shivering is caused by the involuntary contraction of skeletal muscles in response to lower-than-normal

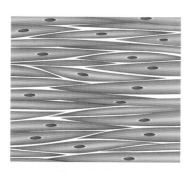

Smooth Muscle

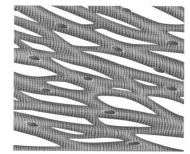

Cardiac Muscle

Skeletal Muscle

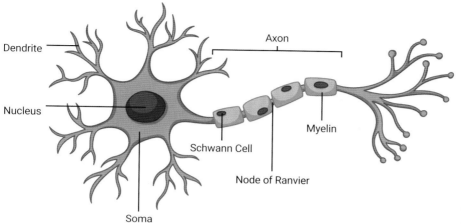

Dendrite

Axon

Nucleus

Myelin

Schwann Cell

Node of Ranvier

Soma

A nerve cell is specialized to transmit information from one part of the body to another, receiving input via its dendrites and transmitting to other nerve and muscle cells via its axon.

body temperature. Skeletal muscle tissue is arranged in bundles surrounded by connective tissue and has a striated appearance when viewed under the microscope. These striations are due to an alternating arrangement of the contractile proteins actin and myosin, together with structural proteins that link the contractile proteins to connective tissues.

Cardiac muscle forms the walls of the heart. It consists of cells known as cardiomyocytes, which also appear striated under the microscope. The cardiac muscle pumps blood through the body and is not under voluntary control. Unlike the cells in skeletal muscle fibre, cardiomyocytes contract without any external stimulation. They are attached to one another through specialized cell junctions called intercalated discs, forming long, branching cardiac muscle fibres that allow the cells to synchronize their actions.

Smooth muscle is responsible for involuntary movements in the internal organs. Smooth muscle cells are spindle-shaped with no visible striations. Smooth muscle contractions move material through the digestive, urinary and reproductive systems as well as the airways and arteries.

Nervous Tissue

Nervous tissue is specialized for the generation and transmission of nerve impulses by cells called neurons. It's made up of these neurons along with glial cells, or neuroglia, which provide support for the neurons. Nervous tissue makes up the central nervous system of the brain and spinal cord along with the network of nerves that runs throughout the rest of the body, the peripheral nervous system.

A neuron has several specialized parts. The cell body, or soma, contains the neuron's organelles, including the nucleus. Dendrites extending from the cell body collect incoming nerve impulses from other neurons, and an axon carries nerve impulses away from the soma to the next neuron in the chain. The axon is encased by a myelin sheath which insulates it, improving transmission of the impulse. Axon terminals are the physical contact with the dendrites of neighbouring neurons.

Neuroglia, which are much smaller than the neurons, outnumber them by as much as 50 to 1. They perform tasks such as bringing nutrients to the neurons, taking away their wastes, and building the myelin sheaths.

Tissue Membranes

Tissue membranes are formed from a thin layer or sheet of cells covering either the outside of the body (skin is a tissue membrane), or lining joints, internal body cavities and vessels. The two basic types of tissue membrane are connective tissue membranes and epithelial membranes.

Connective tissue membranes are formed entirely of connective tissue and are found enclosing organs, such as the kidney, or lining the cavity of a joint. A joint-lining membrane is known as a synovial membrane. Cells in the inner layer of the synovial membrane release synovial fluid, a natural lubricant that reduces friction and enables the bones of a joint to move freely.

Epithelial membranes consist of an epithelial layer attached to a layer of connective tissue. A mucous membrane, also known as a mucosa, lines a body cavity or hollow passageway that is open to the external environment. Mucus, produced by glandular tissue, coats the epithelial layer, which is supported by the underlying connective tissue. These membranes are found lining portions of the reproductive, respiratory, digestive and excretory systems.

Cavities of the body that are not open to the external environment are lined by a serous membrane. Serous fluid secreted by the cells of the epithelium lubricates the membrane and reduces friction and wear between organs. Two serous membranes, called the pleura, cover the lungs and one, called the pericardium, covers the heart. A fourth serous membrane, the peritoneum, lines the peritoneal cavity, covering the abdominal organs and holding many of the digestive organs in place.

Skin is a cutaneous membrane, a multi-layered membrane composed of epithelial and connective tissues. The surface of this membrane exposed to the external environment is covered with dead, hardened cells that help protect the body from losing moisture and act as a barrier to harmful organisms and toxins.

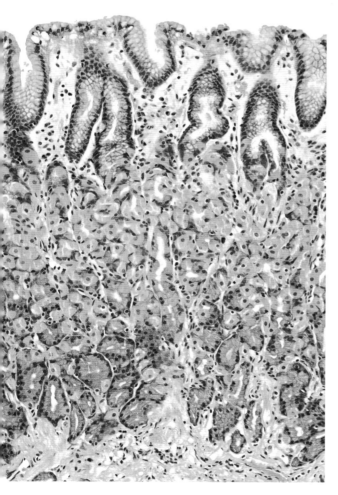

A microscopic cross-section of the gastric mucosa, the mucous membrane that lines the stomach. Around 1 mm thick, it is covered by a thick layer of mucus that protects the stomach from the corrosive effects of gastric juice.

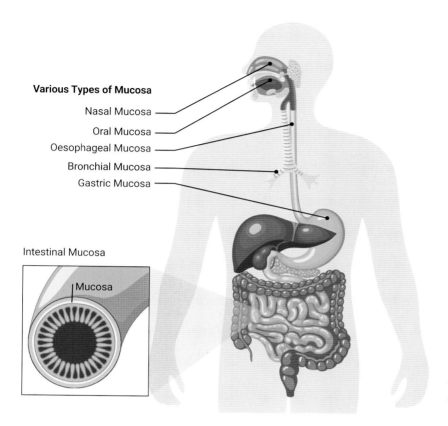

Various Types of Mucosa

Nasal Mucosa

Oral Mucosa

Oesophageal Mucosa

Bronchial Mucosa

Gastric Mucosa

Intestinal Mucosa

Mucosa

Left: Mucous membranes line various cavities in the body that are open to the external environment.

Below: The organs of the body are separated by a number of fluid-filled cavities. Each lung is held within a pleural cavity, separated by two membranes, the visceral pleura and the parietal pleura.

ORGANS AND ORGAN SYSTEMS

The organs of the body are anatomically distinct structures composed of two or more tissue types which perform one or more specific physiological function, for example the cardiac muscle, connective tissue and nerve tissue that make up the heart. A group of organs working together to perform major functions or meeting the physiological needs of the body form an organ system, such as the stomach, intestines and other organs that make up the digestive system. Five major organs – the heart, brain, lungs, liver and kidneys – are considered vital for survival. Failure of any one of these will result in death (in the

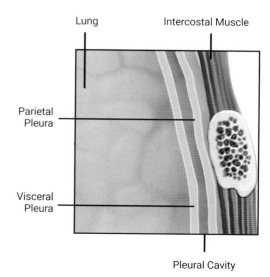

Lung

Intercostal Muscle

Parietal Pleura

Visceral Pleura

Pleural Cavity

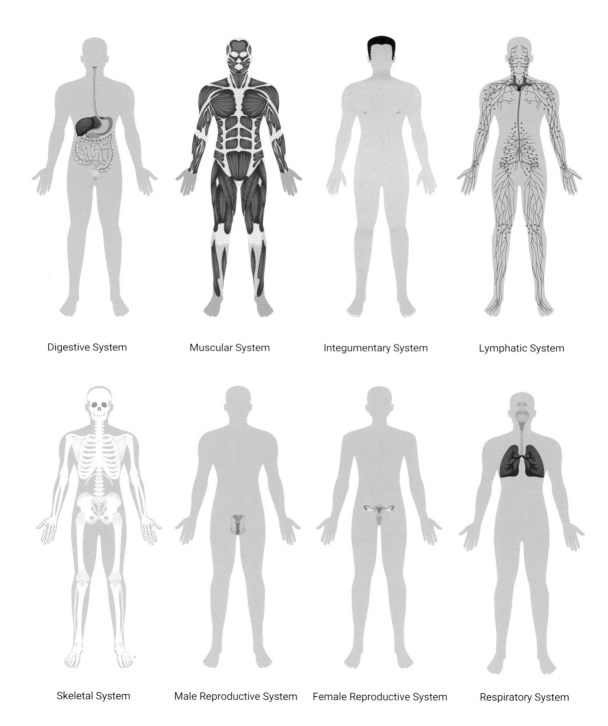

Digestive System

Muscular System

Integumentary System

Lymphatic System

Skeletal System

Male Reproductive System

Female Reproductive System

Respiratory System

case of the paired lungs and kidneys, survival is possible with just one of the pair functioning).

At the top level of the hierarchy of the living being, several organ systems work in harmony to ensure the survival of the organism. For example, processing the food we eat doesn't just involve the digestive system – the cardiovascular and nervous systems also play a role.

The main organ systems at work in the human body, which will be explored in depth in subsequent chapters, are:

Integumentary System
This includes the skin, hair and nails. Its function is to protect the internal organs, providing a barrier against pathogens. It also plays a role in regulating body temperature and in eliminating waste through sweat glands.

Skeletal System
This consists of the teeth, joints and bones, which are connected by tendons, ligaments and cartilage. The skeletal system supports and gives shape to the body, protects the internal organs, provides a store for minerals such as calcium, and is the site of blood cell production. Along with the muscular system, the skeletal system allows movement.

Muscular System
There are three different types of muscles: skeletal muscles, which are attached to bones by tendons and allow for voluntary movements; smooth muscle, controlling involuntary movements of

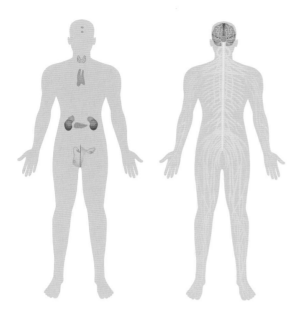

Endocrine System   Nervous System

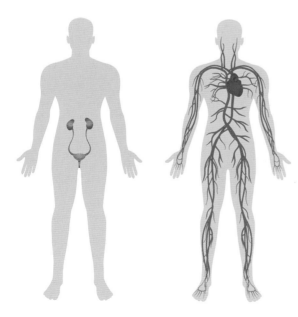

Urinary System   Circulatory System

The human body is composed of a number of interlinked systems that work together to ensure the continuation of life.

internal organs; and cardiac muscle, controlling the involuntary beating of the heart.

Nervous System

This divides into the central nervous system, which includes the brain and spinal cord, and the peripheral nervous system of nerves running throughout the rest of the body. Voluntary and involuntary responses of the body are controlled by the nervous system which also processes information gathered by the sense organs.

Endocrine System

Endocrine hormones are chemical messengers that circulate through the bloodstream. They are produced by glands, such as the pancreas, pituitary and thyroid glands, and control many bodily functions such as metabolism and sexual development.

Cardiovascular System

Also known as the circulatory system, this includes the heart and the blood which it pumps through the veins, arteries and capillaries that make up the blood vessels. The cardiovascular system helps regulate body temperature, transports oxygen and nutrients to the cells of the body and carries away waste products to the lungs and kidneys for elimination.

Respiratory System

Working in tandem with the cardiovascular system, the main function of this system is to provide oxygen to the body and remove waste carbon dioxide by moving air in and out of the lungs and airways.

Digestive System

This system handles the nutritional requirements of the body, breaking down food and liquids through mechanical and chemical means into usable components to provide energy and for repair and growth. Key organs of the digestive system include the stomach, pancreas, liver and intestines.

Urinary System

This system filters and removes excess water, salts and waste materials from the body, primarily through the kidneys, which produce urine for expulsion during urination. The urinary system is involved in regulating blood volume and blood pressure and in the composition of the blood.

Immune and Lymphatic Systems

The immune system is the body's defence against pathogens. The largest organ of the lymphatic system is the spleen, which produces white blood cells to fight infection. The lymphatic system maintains fluid balance in the body, collecting and transporting lymph fluid from the tissues and returning it to the bloodstream. Lymph is rich in nutrients and white blood cells.

Reproductive System

This system is the only body system that is substantially different between male and female. The male system is responsible for producing and transporting gametes (sex cells) to the female reproductive tract which supports the developing offspring until birth.

HOMEOSTASIS

The cells of the human body are bathed in a fluid, called the interstitial fluid, that supplies nutrients and carries away wastes. Survival depends on maintaining the stability of that fluid environment. The processes that together maintain and regulate a stable internal

BODY CAVITIES

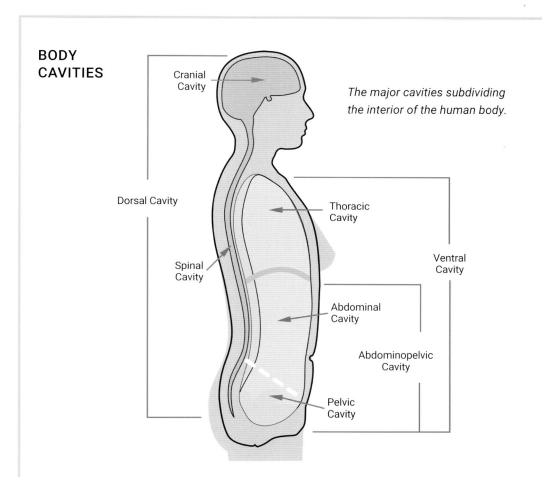

Cranial Cavity

The major cavities subdividing the interior of the human body.

Dorsal Cavity

Thoracic Cavity

Ventral Cavity

Spinal Cavity

Abdominal Cavity

Abdominopelvic Cavity

Pelvic Cavity

The human body is divided into a number of fluid-filled spaces, or cavities, separated from each other by membranes and other structures, that hold and protect the internal organs. The two largest of these cavities are the ventral cavity and the dorsal cavity. The ventral cavity at the front of the trunk holds the lungs, heart, stomach, intestines, and reproductive organs. It is subdivided into the thoracic cavity in the chest, which is further divided into two pleural cavities holding the lungs, and the pericardial cavity holding the heart, and the abdominopelvic cavity in the lower half of the trunk. This is further subdivided into the abdominal cavity holding the digestive organs and the kidneys and the pelvic cavity holding the reproductive organs and organs of excretion. The dorsal cavity is at the back of the body and is subdivided into the cranial cavity containing the brain and the long, narrow spinal cavity inside the vertebral column that contains the spinal cord.

environment within living organisms is called homeostasis. (The prefix – *homeo*, meaning 'similar', rather than *homo*, meaning 'the same' – reflects the fact that the internal conditions remain more or less the same but can vary within certain limits.) Homeostasis explains how an organism can maintain more or less constant internal conditions that allow it to survive in an external environment that is always changing and may sometimes even be hostile.

The structure and function of all animals, not just humans, is geared towards the needs of homeostasis. Individual cells engage in activities that ensure their survival; the cells that make up the tissues in complex organisms like us contribute to the survival of the organism as a whole; and the combined contributions of cells, tissues and organ systems ensure the essential maintenance of a stable internal environment in which the cells can remain healthy.

Homeostasis involves co-ordinating the activities of various systems and organs within the body. Achieving this requires reliable communication between cells and tissues in different parts of the body. Maintaining the stable conditions of homeostasis requires a system with three components – a receptor, a control centre and an effector. These work together in a negative feedback loop which works to oppose or reset the stimulus that triggers action. Sensory receptors are cells that are stimulated by changes in the environment, for example nerve cells that detect variations in temperature or cells in the blood

vessels that detect changes in blood pressure. Triggering these detectors causes them to send a signal to a control centre, which determines an appropriate response.

One of the most important such control centres is the hypothalamus, a region of the brain that oversees everything from body temperature to heart rate, blood pressure, and sleep and wake cycles. The control centre signals an effector which will bring about the changes needed to restore balance. This effector might be the muscles that make you shiver when you are cold, or a gland in the endocrine system that releases a hormone to regulate calcium levels in the blood. Once balance has been restored the sensory cells will signal the change to the control centre which will then deactivate the effectors.

The endocrine system, which produces hormones, is a crucial driver of homeostasis and the regulation of blood sugar levels is a good example of a negative feedback loop. The negative feedback mechanism is triggered by high glucose levels and shut off when they fall to lower levels. The presence of glucose in the bloodstream stimulates the pancreas to produce insulin, which signals the liver to store excess glucose in the form of glycogen. As glucose concentrations in the blood are reduced the pancreas stops producing insulin and the liver stops producing glycogen. Glucose levels in the blood are thus maintained within the range necessary for the body's requirements.

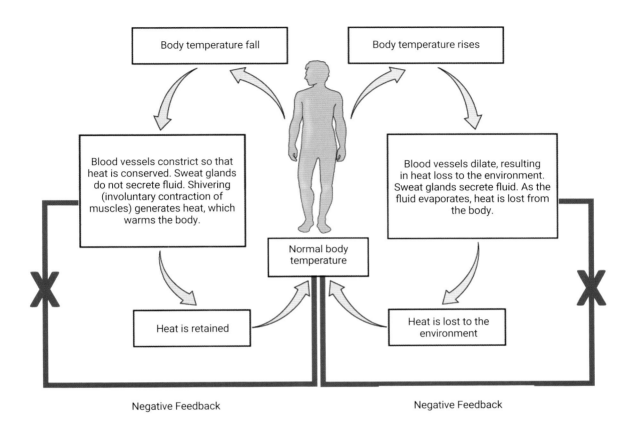

The homeostasic equilibrium of the human body is maintained by a number of negative feedback loops, triggering a response that aims to oppose any imbalance. For example, body temperature is regulated by involuntary actions such as constricting or dilating blood vessels.

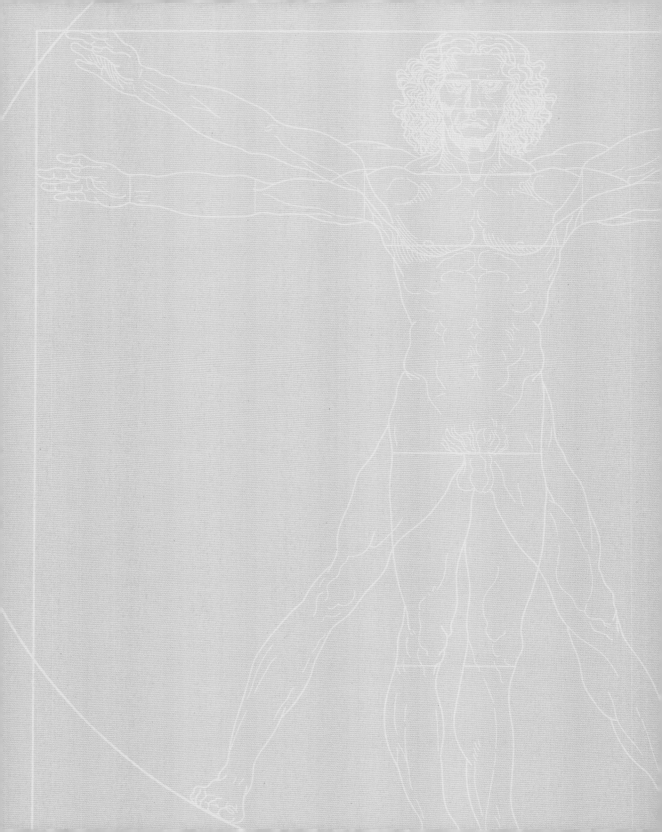

CHAPTER TWO

The Molecules of Life

CHAPTER TWO

The Molecules of Life

Our lives depend on trillions of chemical reactions taking place simultaneously within the cells that make up our bodies. Each cell is like a chemical factory, breaking down and reassembling raw materials and producing wastes that need to be disposed of. The sum of all these life-sustaining chemical reactions is called metabolism. Certain molecules are characteristic of life – only living cells synthesize carbohydrates, proteins, lipids and nucleic acids. These macromolecules are large carbon-based structures called polymers, assembled from smaller units, called monomers. Think of each monomer as an individual bead on a polymer necklace. All of the millions of

Carbohydrates are formed from carbon, hydrogen and oxygen. They range in complexity from energy-providing simple sugars, such as glucose, to large, complex structural polysaccharides, such as cellulose.

SUGARS

Glucose

Sucrose (Table Sugar)

● Carbon   ● Hydrogen   ● Oxygen

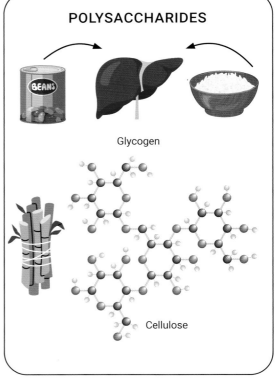

POLYSACCHARIDES

Glycogen

Cellulose

biochemical compounds found in living things are formed from carbon, hydrogen and oxygen, along with nitrogen, phosphorus, sulphur and other elements.

CARBOHYDRATES

Carbohydrates are the most abundant of the biological molecules, used by cells as an energy source and as structural materials. Most are formed from carbon, hydrogen and oxygen. The simplest forms of carbohydrate are the simple sugars, or monosaccharides, the commonest of which is glucose. Plants synthesize glucose using carbon dioxide and water through the process of photosynthesis. Galactose (part of lactose, or milk sugar) and fructose (found in fruit) have the same chemical formula ($C_6H_{12}O_6$) as glucose but differ structurally and chemically because the atoms in their carbon chains are differently arranged.

Disaccharides form when two monosaccharides link together. Common disaccharides include lactose, maltose and sucrose. Lactose, formed from glucose and galactose, is found naturally in milk. Maltose, or malt sugar, is formed from two glucose molecules. Sucrose, commonly known as table sugar, is the most common disaccharide and is formed from glucose and fructose.

A *polysaccharide* is a long chain of monosaccharides. The chain may be branched or unbranched, and it may be formed of different types of monosaccharides. Polysaccharides are often very large molecules with thousands of monosaccharides bonded together. Starch, glycogen, cellulose and chitin, which forms the exoskeletons of insects and other animals, are all examples of polysaccharides.

Starch is the stored form of sugars in plants and is made up of amylose and amylopectin, which are both polymers of glucose. The starch that is consumed by animals is broken back down by the digestive system into smaller sugar molecules, such as glucose, to provide cells with energy. Cellulose, one of the most abundant of the polysaccharides, is a major constituent of plant cell walls. Although indigestible by humans, cellulose, commonly referred to as fibre, is an important part of the diet, helping to keep the digestive tract healthy and believed to reduce the risk of heart disease.

Glucose is stored in the body in the form of glycogen which is made up of monomers of glucose. Glycogen is a highly branched molecule usually stored in the liver and muscle cells. It provides a reserve of energy for the body that can be accessed quickly when needed. When blood sugar levels decrease, or during a period of activity, glycogen is broken down to release glucose.

Grains, fruits and vegetables provide most of the carbohydrate in the human diet. All of the body's cells use glucose as an energy source. Although most cells can also break down other organic compounds for fuel, red blood cells and nerve cells in the brain, spinal cord and peripheral nervous system can only use glucose for fuel. Carbohydrates also play a role in cell structure. Some bind with proteins to produce glycoproteins, and others with lipids to form glycolipids, both of which are found in the cell membrane.

LIPIDS

Lipids, which include fats, oils, waxes and steroids, perform many different functions in the cell. Lipids are the building blocks of many hormones and are an important constituent of the plasma membrane that surrounds the cell. Lipids provide insulation from the environment and act as an energy store in cells.

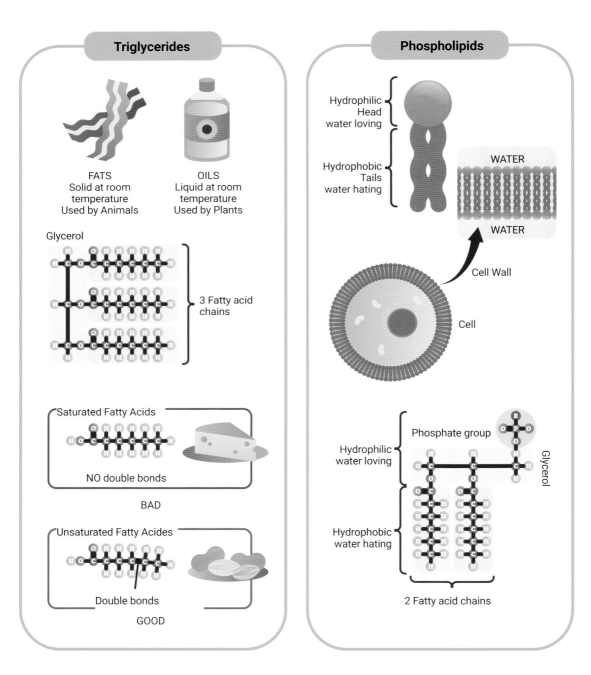

Triglycerides

FATS
Solid at room
temperature
Used by Animals

OILS
Liquid at room
temperature
Used by Plants

Glycerol

3 Fatty acid
chains

Saturated Fatty Acids

NO double bonds

BAD

Unsaturated Fatty Acides

Double bonds

GOOD

Phospholipids

Hydrophilic
Head
water loving

Hydrophobic
Tails
water hating

WATER

WATER

Cell Wall

Cell

Phosphate group

Hydrophilic
water loving

Hydrophobic
water hating

Glycerol

2 Fatty acid chains

Lipids in their different forms have many vital roles to play in the human body, from forming the membranes that surround individual cells, to acting as hormones and storing energy in the form of body fat.

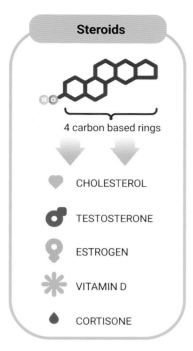

Steroids

4 carbon based rings

CHOLESTEROL

TESTOSTERONE

ESTROGEN

VITAMIN D

CORTISONE

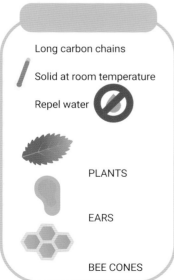

Long carbon chains

Solid at room temperature

Repel water

PLANTS

EARS

BEE CONES

Lipids are polymers, made up of linked units called fatty acids. There are two types of fatty acids – saturated fatty acids and unsaturated fatty acids. These differ according to the number of bonds between their constituent carbon atoms and the number of hydrogen atoms they contain.

Saturated fatty acids form straight chains with single bonds between their carbon atoms and as many hydrogen atoms as possible attached. Because the straight chains can be packed tightly together, saturated fatty acids are a good compact energy store. They have relatively high melting points and are usually solid at room temperature. Butter is a typical example of a saturated fat.

Unsaturated fatty acids have double or triple bonds between the carbon atoms and fewer hydrogen atoms attached. Monounsaturated fatty acids have one less hydrogen atom than the equivalent saturated fatty acid while polyunsaturated fatty acids have at least two fewer hydrogen atoms. The chains they form are bent and so can't be packed closely together. They have relatively low melting points and are liquid at room temperature. Plants use unsaturated fatty acids as an energy store. Olive and peanut oils are a source of monounsaturated fats, while sunflower oil and many nuts and seeds provide a source of polyunsaturated fats.

Lipids may also include other chemical components along with the fatty acid molecules. Triglycerides combine three fatty acid molecules with a molecule of glycerol, a sweet-tasting colourless compound that is also known as a sugar alcohol. Triglycerides carry twice as much energy as carbohydrates and are stored in the body's fat cells. Triglycerides are the main constituent of human body fat. Hormones trigger the release of triglycerides back into the bloodstream to meet the body's energy requirements.

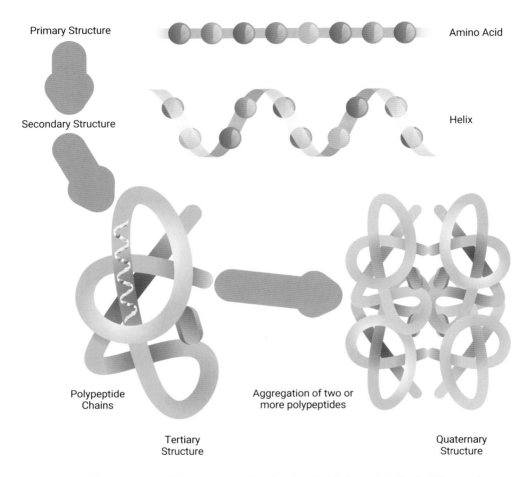

Primary Structure

Amino Acid

Secondary Structure

Helix

Polypeptide
Chains

Aggregation of two or
more polypeptides

Tertiary
Structure

Quaternary
Structure

Proteins are a diverse group of molecules that take part in the building and running of the human body at every level. Formed from chains of smaller amino acids, they twist and bend into shapes of great complexity that allow them to fulfil a huge variety of functions.

A major component of cell membranes are phospholipids. These molecules have a 'head' consisting of a glycerol molecule and a phosphate group, to which is attached a 'tail' formed from two long, fatty acids. The hydrophilic head is attracted to water, whereas the hydrophobic tail is repelled by it. Many phospholipids line up tail to tail to form a bilayer, a two-layered cell membrane, with the hydrophilic heads facing the watery interior of the cell in the inner layer and the similarly watery space around the cell in the outer layer.

Steroids are vital biochemicals. Many are hormones, including the sex hormones such as oestrogen and testosterone, while others, such as cholesterol, are important components of cell membranes. Steroids are lipids with a ring structure. Each has a core of 17 carbon atoms arranged in four rings of five or six carbon atoms apiece. The function of the steroid depends on which other components are attached to this inner core.

PROTEINS

Proteins are formed from long chains of smaller molecules called amino acids. A short chain of amino acids is called a peptide and longer chains are called polypeptides. One or more polypeptides together form a protein. Large proteins may consist of thousands of amino acids chained together. Proteins are an incredibly diverse group of molecules with structures varying on multiple levels. This diversity is what allows proteins to fulfil such a variety of important roles. There are at least 10,000 different proteins at work in the human body.

Many proteins are enzymes, biological catalysts that control the speed of chemical reactions in cells. Enzymes are highly specific in their actions and many thousands of biochemical processes are each controlled by one particular enzyme. Other proteins perform structural functions in the cell and in the body as a whole. Proteins are found in skin, bone, muscle and almost every other body part and tissue. Antibodies, a vital component of the immune system, are proteins, binding to foreign proteins from invading microorganisms and marking them for attack. The ability of proteins to bind with other molecules allows haemoglobin, a protein in red blood cells, to act as a carrier for oxygen, transporting it around the body.

The human body cannot itself synthesize all of the amino acids it needs and must obtain some from food. During digestion the proteins we consume are broken down into their constituent amino acids to be reassembled into our own proteins. The process of protein synthesis is controlled by genes.

NUCLEIC ACIDS

Nucleic acids are the biochemicals responsible for heredity, passing information from one generation to the next, and for acting on that information to synthesize proteins in the cell. They are built from smaller subunits called nucleotides which bind together in chains to form polynucleotides. The nucleic acid deoxyribonucleic acid, or DNA, consists of two polynucleotide strands. Ribonucleic acid, RNA, has a single polynucleotide strand. Each nucleotide itself is built from smaller molecules: a sugar molecule (deoxyribose in DNA and ribose in RNA); a phosphate group; and a nitrogen base. The sugar and phosphate groups form the backbone of the polypeptide chain. There are four different nitrogen bases – cytosine, thymine, adenine and guanine (in RNA thymine is replaced by uracil). The twin strands of DNA are held together by bonds between the nitrogen bases – cytosine always binds with guanine and adenine always binds with thymine – winding the molecule into its characteristic double helix shape.

The sequence of bases in DNA is the foundation of the genetic code; a gene is basically a length of DNA, carrying the instructions for the assembly of proteins. RNA translates this information to link amino acids in the correct order to form the relevant protein.

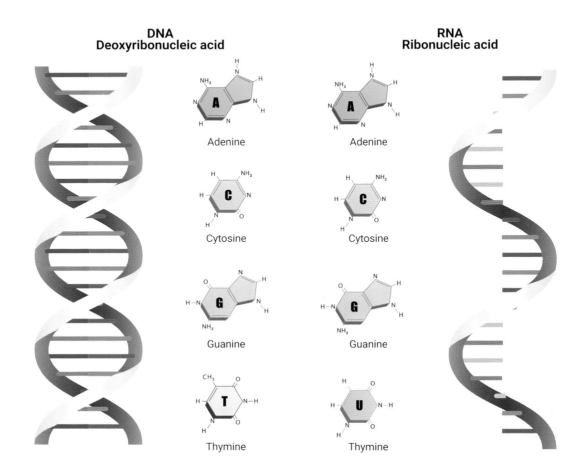

DNA
Deoxyribonucleic acid

RNA
Ribonucleic acid

Adenine

Adenine

Cytosine

Cytosine

Guanine

Guanine

Thymine

Thymine

*Nucleic acids in the form of DNA, which carries the genetic code, and RNA,
which translates that code, are the molecules responsible for passing hereditary
information from one generation to the next. Each is formed from four different
bases joined along a backbone of sugar and phosphate molecules.*

ATP

There is a third type of nucleic acid that exists only as a single unit and has no role in the coding or assembly of proteins. Nonetheless, it is one of the most important of the biochemicals. Adenosine triphosphate, ATP, is an energy-carrying molecule that is used to power most of the processes in the cell. Glucose, the ultimate source of energy in the body, is easily transported through the bloodstream to the cells but in order to be useful to the cell its energy has to be parcelled out into more manageable amounts.

ATP consists of a sugar, adenosine and

three phosphate groups. The bond between the second and third phosphate groups holds energy that can be utilized by the cell when the bond is broken. In the process ATP becomes ADP, adenosine diphosphate. The ADP molecules can be recharged by reattaching the phosphate group to make ATP again. This is accomplished by using the energy of glucose to produce molecules of ATP – under nominal conditions each glucose molecule can produce 38 molecules of ATP. This process, called cellular respiration, takes place in the cell's mitochondria. In a single second, the average human body cell will use the energy from around 10 million ATP molecules.

Adenosine triphosphate (ATP) is also a nucleic acid but not one that forms a chain. Its function is to carry and release the energy the cells of the body need to carry out their functions.

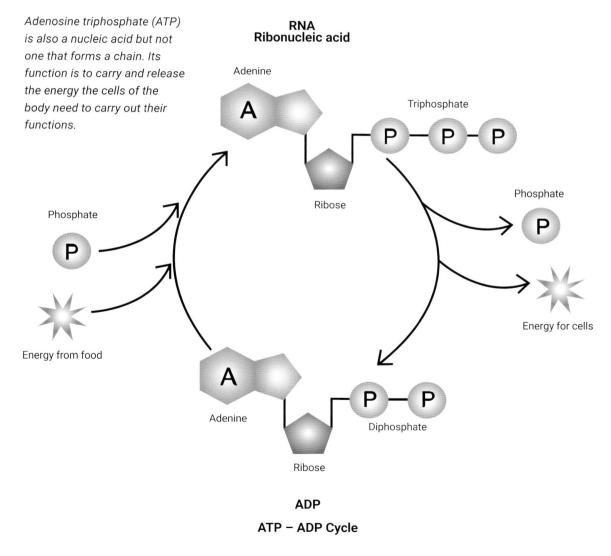

RNA
Ribonucleic acid

Adenine

Triphosphate

Ribose

Phosphate

Phosphate

Energy from food

Energy for cells

Adenine

Diphosphate

Ribose

ADP

ATP – ADP Cycle

The Integumentary System

The Integumentary System

The human body is protected by an outer layer, or integument, commonly referred to as skin, a tough, pliable barrier against infection and other hazards of the environment. It is the largest organ system in the human body, covering an area of 1.5 to 2 m² (16–21 ft²) and accounting for around 16 per cent of the body weight of the average adult. Most areas of the skin are about as thick as a paper towel, but it is thicker in areas such as the soles of the feet that are subject to greater wear.

The skin is the major organ of the integumentary system, which also includes hair and nails. The integumentary system fulfils several roles in the human body:

Physical protection: the tough outermost layer of the skin, the epidermis, can withstand wear and tear and nerves in the dermis beneath warn of possible damage; fatty adipose tissue in the hypodermis, the innermost layer of skin, provides physical cushioning from injury.
Immunity: the skin is the first line of defence against infection, providing both a physical and

The complexity of the body's outer covering of skin reflects the variety of essential functions it carries out, from protecting the body's internal organs, to regulating its temperature and providing sensory information about the environment.

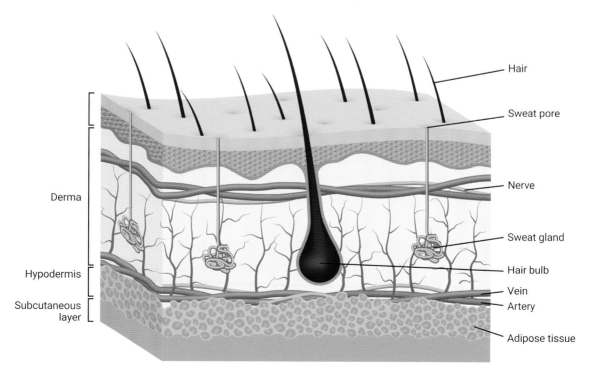

Derma

Hypodermis

Subcutaneous layer

Hair

Sweat pore

Nerve

Sweat gland

Hair bulb

Vein

Artery

Adipose tissue

an antimicrobial barrier preventing disease-causing microorganisms from entering the body.
Wound healing: when the body is injured, the integumentary system plays an important role in the wound healing process.
Sensation: nerve receptors in the skin which detect temperature, touch, pain and vibration, are one of the main ways in which we interact with the environment.
Thermoregulation: the skin's large surface area and many blood vessels allow it to conserve and release heat through vasoconstriction and vasodilation.
Vitamin D synthesis: the body's primary source of vitamin D, which is crucial for bone health, is through exposure of the skin to sunlight.

THE SKIN

The skin is a complex organ with blood vessels, hundreds of glands and nerve endings, and thousands of pigment-producing cells packed into every square centimetre of a layer of tissue that is on average just 2 mm thick. The skin has three main layers. The outermost layer, the epidermis, provides a waterproof barrier and contributes to skin tone. The dermis, located beneath the epidermis, contains connective tissue, hair follicles, blood vessels, lymphatic vessels and sweat glands. The hypodermis, the deeper subcutaneous tissue, consists of fat and connective tissue. It anchors the skin to the underlying body parts but still allows it some flexibility and movement. The fat in the hypodermis helps insulate and cushion the body.

The Epidermis

The epidermis, the thinnest of the three main skin layers, is just 0.05 mm thick on the eyelids and around 1.5 mm thick on the soles of the feet and

SKIN COLOUR

Around 8 per cent of the epidermal cells are melanocytes, producing the pigment melanin which protects deeper layers of the skin from ultraviolet light. Melanin is the main substance determining the colour of the skin. Skin colour is not due to the number of melanocytes present but to how active they are, with light-skinned people producing less melanin. Two other substances, carotene and haemoglobin, also contribute to skin colour, especially in light-skinned people. The pigment carotene in the epidermis gives skin a yellowish tint, especially where there are low levels of melanin. Haemoglobin, the red pigment found in red blood cells, is visible as a pinkish tint, mainly in skin with low levels of melanin.

palms of the hands. There are no blood vessels in the epidermis, requiring the cells to absorb oxygen directly from the air and nutrients by diffusion from the cells of the dermis. A variety of cells are found in the epidermis, mainly keratin-producing cells called keratinocytes, or squamous cells, which account for 90 per cent of the epidermis.

Langerhans cells, which are part of the immune system protecting against pathogens, make up about 1 per cent of the epidermal cells. Fewer than 1 per cent are Merkel cells which connect to nerve endings in the dermis and respond to a light touch. The epidermis can itself be divided into five further layers. Beginning with the deepest layer of the epidermis these are:

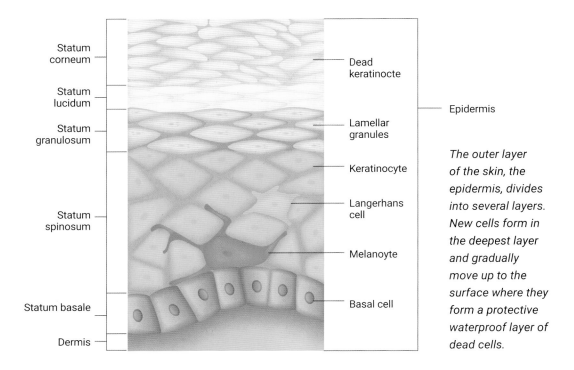

Statum corneum

Statum lucidum

Statum granulosum

Statum spinosum

Statum basale

Dermis

Dead keratinocte

Lamellar granules

Keratinocyte

Langerhans cell

Melanoyte

Basal cell

Epidermis

The outer layer of the skin, the epidermis, divides into several layers. New cells form in the deepest layer and gradually move up to the surface where they form a protective waterproof layer of dead cells.

Stratum basale

New skin cells develop in this layer from stem cells called basal cells which give rise to the keratinocytes, which are pushed up as more cells form beneath them. The stratum basale is also where the melanocytes and Merkel cells are located.

Stratum spinosum

The thickest of the epidermal layers, consisting mostly of keratinocytes bound together by sticky proteins called desmosomes; it gives the epidermis strength and flexibility. This layer also contains the Langerhans cells, which attach themselves to pathogens such as bacteria and other substances that invade damaged skin and trigger a response from the immune system.

Stratum granulosum

A thin layer of cells in which the keratinocytes take on a granular appearance. The cells in this layer are now too far from the dermis to obtain nutrients and begin to die, leaving a tough keratin-filled shell.

Stratum lucidum

A thin, transparent layer in which the keratinocytes take on a flatter shape. This is only really evident where the skin is thickest, such as on the soles of the feet and palms of the hands.

Stratum corneum

The outermost, visible layer of the epidermis. Here, a layer of tightly packed dead keratinocytes provides a strong barrier giving protection from pathogens, ultraviolet radiation, heat and physical

A microscopic cross-section through the skin shows the finger-like projections that extend from the papillary dermis into the epidermis, supplying it with nutrients. The reticular dermis beneath provides strength and structure and houses the sweat glands.

damage. These dead cells are continually shed and replaced as new keratinocytes grow in the stratum basale and are pushed up to the surface, a process that takes around 48 days. Fats in the stratum corneum prevent water being easily lost or absorbed through the skin.

The Dermis

The dermis lies beneath the epidermis and is separated from it by the basement membrane. It is the thickest of the skin's three main layers, and itself consists of two sublayers, the papillary layer just below the basement membrane with the reticular layer beneath that. The dermis is held together by collagen, a tough, insoluble protein found throughout the body in the connective tissues. It is made by fibroblasts, skin cells that give the skin its strength and resilience. Elastin, a similar protein, gives the skin flexibility and allows it to spring back into place when stretched.

Papillary dermis

This supplies nutrients to the epidermis located above it through an extensive system of tiny blood vessels, within finger-like projections called papillae which extend into the epidermis. The blood vessels in the papillae also carry away wastes and transport vitamin D from the skin to the rest of the body. Constricting or expanding these blood vessels helps regulate body heat. The interlocking nature of the papillae with the epidermis strengthens the connection

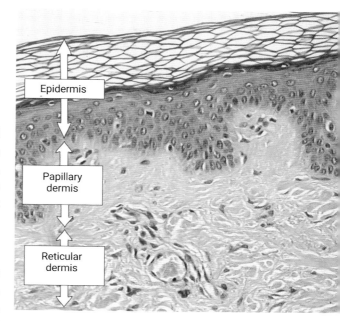

between the two layers. The epidermal ridges that arise on the fingers are what we commonly know as fingerprints. Because these patterns are genetically determined, fingerprints were commonly used as a means of identification, although this method is now being superseded by DNA analysis.

Reticular dermis

The thickest and lowest of the sublayers, this is formed from thick collagen and elastin fibres arranged parallel to the surface of the skin. It provides structure, strength and elasticity as well as housing most of the structures in the dermis, such as hair follicles, sebaceous glands and sweat glands.

Sweat glands

Sweat is a fluid containing mainly water and salts. The sweat glands that produce it have ducts

Fingerprints, which are unique to every individual, arise from the projections of the papillary dermis in to the epidermis above.

an oily sweat, which empties out through the ducts of the apocrine glands into hair follicles. The digestion of apocrine sweat by bacteria living on the skin is what causes body odour.

Sebaceous glands

Sebaceous glands produce a thick, fatty substance called sebum, which is secreted into hair follicles and makes its way to the skin surface. It provides a waterproof coating for the hair and skin, helping to prevent them from drying out. Sebum also inhibits the growth of microorganisms on the skin. Sebaceous glands are found in every part of the skin with the exception of those places where hairs does not grow, i.e. the palms of the hands and soles of the feet.

that carry the sweat to hair follicles, or to the surface of the skin. There are two different types of sweat glands: eccrine glands and apocrine glands. Eccrine glands are found in skin all over the body, their ducts emptying on to the skin surface through pores. These sweat glands are involved in temperature regulation. Apocrine sweat glands are larger and are found only in the skin of the armpits and groin. They are inactive until puberty, at which time they begin to produce

Hair

Hair grows from hair follicles in the dermis of the skin. With few exceptions, such as the palms

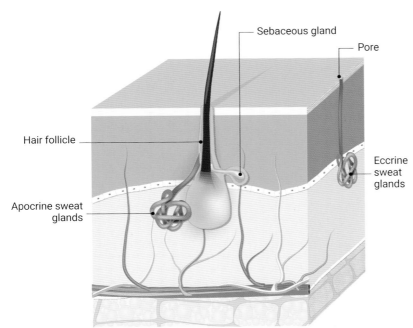

Sebaceous gland

Pore

Hair follicle

Eccrine sweat glands

Apocrine sweat glands

Apocrine sweat glands, located in the skin of the armpit and groin areas, only become active after puberty and empty out on to hair follicles. Eccrine glands are found all over the body and empty through pores directly on to the skin surface.

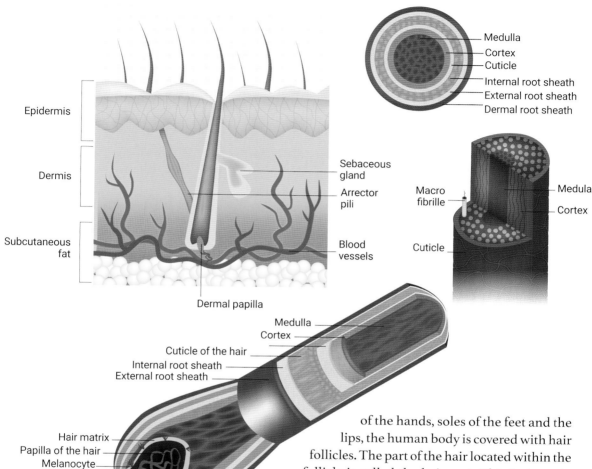

Medulla
Cortex
Cuticle
Internal root sheath
External root sheath
Dermal root sheath

Epidermis

Dermis

Subcutaneous fat

Sebaceous gland

Arrector pili

Blood vessels

Dermal papilla

Macro fibrille

Medula

Cortex

Cuticle

Medulla
Cortex
Cuticle of the hair
Internal root sheath
External root sheath

Hair matrix
Papilla of the hair
Melanocyte

Blood vessels

Above: Hairs grow from follicles embedded in the dermis and extend out beyond the surface of the skin. Each has an associated sebaceous gland, lubricating and waterproofing it, and an arrector pili muscle that can pull it erect. Structurally, each hair divides into cuticle, cortex and medulla zones.

of the hands, soles of the feet and the lips, the human body is covered with hair follicles. The part of the hair located within the follicle is called the hair root. This is the only living part of the hair and is where hair growth begins. The part of the hair that is visible above the surface of the skin is called the hair shaft and is considered dead.

The hair shaft is a hard filament, mostly formed from keratin, and may grow very long. Hair normally grows in length by over 1 cm (0.4 in) a month.

In cross-section, a hair shaft divides into three zones: the cuticle, cortex and medulla. The cuticle (or outer coat) consists of several layers of flat, thin, overlapping cells like the tiles on a roof, called keratinocytes. The cuticle is covered with a layer of lipids, just one molecule thick which, along with the overlapping structure, helps it to repel water. The middle zone of the shaft, the cortex, consists of rod-like keratin bundles, which give hair its mechanical strength. The cortex also contains melanin, which gives hair its colour. The innermost part of the hair shaft, the medulla, is not always present, but where it is, it contains highly pigmented cells full of keratin.

Hair colour is produced by melanin, which is created in the hair follicles. Two forms of melanin are found in human hair: eumelanin, the dominant pigment in brown and black hair and pheomelanin, the dominant pigment in red hair. Blond hair occurs when there is only a small amount of melanin present. When melanin production slows down and eventually stops as we get older, the result is grey and white hair.

The consistency and curliness of hair is determined by the shape of the hair shaft. Round hair shafts produce straight hair; oval or other shaped hair shafts produce wavy or curly hair. Fine hair has shafts with the smallest circumference, and coarse hair the largest.

Hair serves a variety of functions, including thermoregulation, protection and sensory input. Each hair follicle is connected to a smooth muscle called the arrector pili that contracts in response to nerve signals from the sympathetic nervous system. This causes the hair shaft to stand up, a phenomenon visible as 'goosebumps', trapping an insulating layer of air that helps prevent heat loss. Hair on the head protects the skull from the sun, while hair in the nose and ears and around the eyes, traps dust particles and microbes. The eyebrows prevent sweat and rain from dripping into the eyes. Nerve endings surrounding the base of each hair follicle make hair extremely sensitive to air movement or other disturbances in the environment, as well as detecting the presence of insects or other potentially damaging intruders on the skin surface.

Nails

Fingernails and toenails protect the soft tissues of the fingers and toes from injury. The counter pressure exerted on the tissue of the finger by the fingernails enhances sensation, allowing precise movements of the fingertips and the ability to pick up tiny objects. Fingernails can also be used to a certain extent for cutting and scraping. Toenails help to prevent injuries and infections and also exert counter pressure on the toes that help with balance while walking. People who have lost toenails often have problems with walking.

The main parts of the nail include the nail root, the portion of the nail found under the surface of the skin at the near end of the nail. Surrounding the nail root is a deep layer of living epidermal tissue, known as the nail matrix, which contains stem cells that divide to form the keratinocytes that make up the nail. The nail root slowly grows out of the skin, becoming the nail plate, the visible part of the nail, formed from tough, hard, translucent sheets of dead cells filled with keratin, which gives the nail hardness and flexibility. As the nail grows longer, the cells of the nail root and nail plate are pushed towards the end of the finger or toe as new cells are formed in the nail matrix.

Nails grow at an average rate of around 3 mm a month, the actual rate depending on factors such as age and diet. Fingernails grow up to four times faster than toenails. If a fingernail is lost, it takes between three and six months to regrow completely. A toenail will take 12 to 18 months.

The area of skin under the nail plate is the nail bed. A layer of dead epithelial cells overlapping

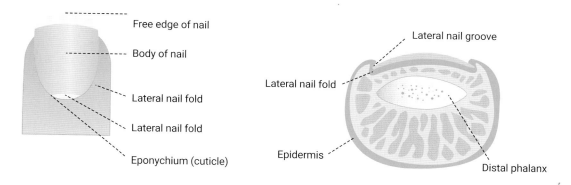

Free edge of nail

Body of nail

Lateral nail fold

Lateral nail fold

Eponychium (cuticle)

Lateral nail groove

Lateral nail fold

Epidermis

Distal phalanx

The external, visible part of a nail is shown on the top left. At top right, a cross section shows the phalanx underlying the nail, while more of the internal structure of the nail, including the nail matrix from which it grows, is shown beneath.

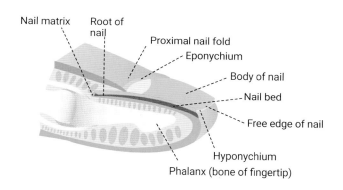

Nail matrix

Root of nail

Proximal nail fold

Eponychium

Body of nail

Nail bed

Free edge of nail

Hyponychium

Phalanx (bone of fingertip)

the edge of the nail plate form the cuticle which helps seal the edge of the nail and prevents infection of the tissues beneath. The groove in the skin in which the side edges of the nail are embedded is called the nail fold.

At the end of the nail plate near the root there is a whitish crescent shape called the lunula. This is where a part of the nail matrix is visible through the nail plate. The lunula is most obvious in the nails of the thumbs, and may not be visible at all in the nails of the little fingers.

Vitamin D

Vitamin D_3 is essential for good health as it is needed by the digestive system for the absorption of calcium. Lack of the vitamin leads to conditions such as rickets, characterized by bent limbs and weak bones. The human body synthesizes vitamin D_3 through exposure to sunlight, a process that begins with a molecule present in the skin called 7-dehydrocholesterol (7-DHC), which is found to a certain extent throughout the skin but which is most highly concentrated in the stratum basale and stratum spinosum of the epidermis. Ultraviolet light penetrates through the skin, converting 7-DHC into a precursor of vitamin D_3 called $preD_3$. A protein, called vitamin D binding protein, binds to the vitamin D_3 and transports it into the bloodstream. At normal body temperature, the reaction can continue for around three days, even without further exposure to sunlight. Given good health and nutrition, moderate sun exposure produces sufficient vitamin D_3 for the body's needs.

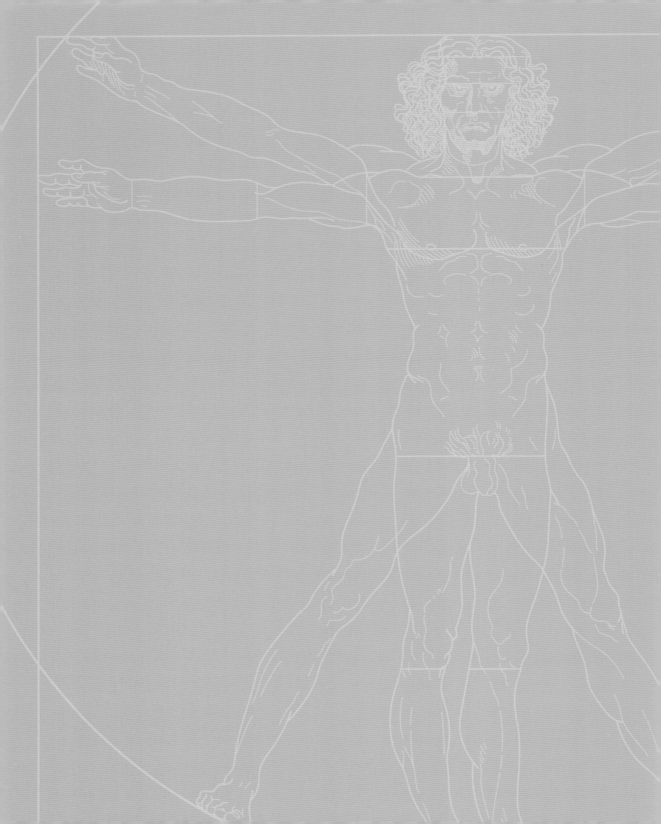

CHAPTER FOUR

The Skeletal System

The Skeletal System

The skeletal system acts as the central framework for the body, working together with the muscular system to provide strength and stability. The human skeleton is formed from bone tissue, a type of connective tissue. Other types of connective tissue, in the form of tendons, ligaments and cartilage, hold the bones of the skeleton together in a complex of over 230 joints. This assemblage of bones and joints makes up the skeletal system, the body's internal framework, providing support and protection for the internal organs and an anchoring point for muscles. Altogether the skeleton makes up around a fifth of a healthy adult's body weight.

An adult human skeleton has 206 bones (at birth we have 300 or so but some fuse together as we grow). The bones of the skeleton range in size from the lentil-sized auditory ossicles deep inside the ear that play an essential role in hearing, to the long, load-bearing femur or thigh

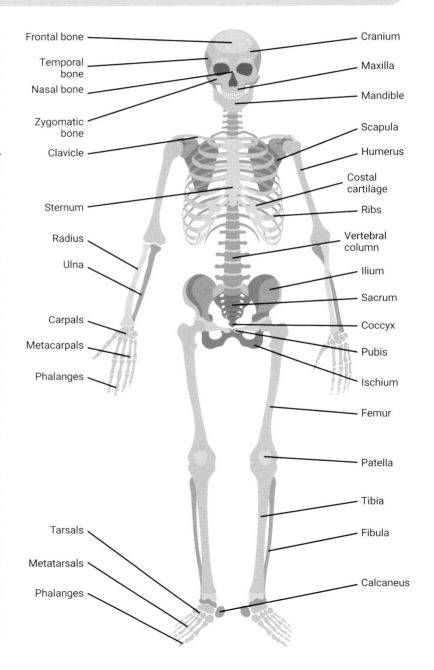

Frontal bone — Cranium
Temporal bone — Maxilla
Nasal bone — Mandible
Zygomatic bone — Scapula
Clavicle — Humerus
— Costal cartilage
Sternum — Ribs
Radius — Vertebral column
Ulna — Ilium
— Sacrum
Carpals — Coccyx
Metacarpals — Pubis
Phalanges — Ischium
— Femur
— Patella
— Tibia
Tarsals — Fibula
Metatarsals — Calcaneus
Phalanges

The major bones of the human skeletal system.

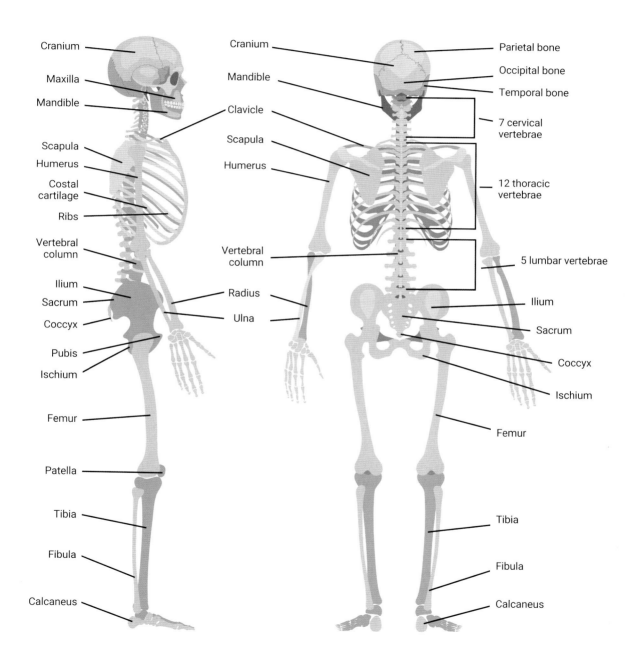

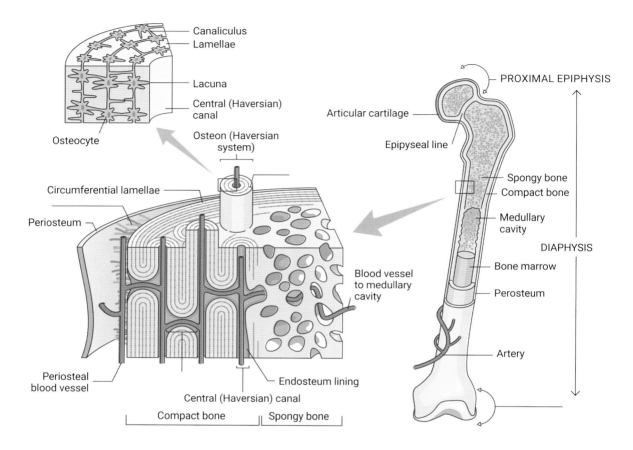

bone, which may account for a quarter of an adult's height. Bones are a storehouse of minerals such as calcium and potassium that are essential for body functions, absorbing and releasing them according to the body's needs. They are also a repository for fat, and some are the production sites for blood cells of all kinds. More than simply playing a supporting role, the skeleton is a vital and complex part of the human body that grows, repairs and renews itself constantly.

Anatomically, the skeleton is divided into the axial skeleton, which contains 80 bones, including the bones of the skull and spine and the ribcage; and the appendicular skeleton made

The outwardly smooth appearance of a bone hides the complex structure within. Blood vessels thread through a network of canals bringing nutrients to, and removing wastes from, the living cells embedded in the bone matrix. The hard outside of the bone is formed from compact bone tissue while less dense, porous, spongy bone tissue fills part of the interior, surrounding a central cavity filled with bone marrow.

up of the remaining 126 bones, including the shoulder bones, the pelvis and the bones of the arms, legs, hands and feet.

BONE TISSUE

Bone, or osseous, tissue is very different from other body tissues. Many of the functions of bone depend on its characteristic hardness. There are two types of bone tissue: compact bone and spongy, or cancellous, bone.

Compact bone is made up of closely packed units called osteons, or haversian systems. A single osteon consists of a central haversian canal surrounded by concentric rings of matrix, between which bone cells called osteocytes are located in spaces called lacunae. The haversian canal carries blood vessels that radiate out through small channels in the matrix.

The matrix is made up of around one-third collagen fibres to two-thirds calcium phosphate salt. The salt crystals give bones their hardness and strength, while the collagen fibres give the bone some flexibility. Without the collagen, bone would be too brittle; without the salts it would be too flexible and unable to bear much weight. *Spongy bone* is lighter and less dense than compact bone. It consists of plates and bars of bone next to small, irregular cavities containing red bone marrow. The plates are organized to provide lightness and strength. The blood supply to the bone comes through small channels connecting the cavities, rather than via a central haversian canal.

Bone cells

Less than a fiftieth of the mass of bone is made up of bone cells but they are vital for it to function effectively. There are three types of bone tissue cells: osteoblasts, osteocytes and osteoclasts.

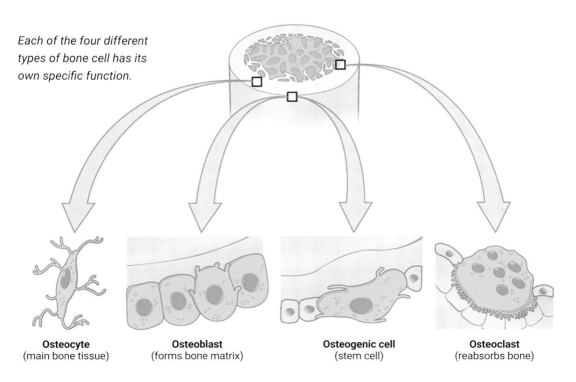

Each of the four different types of bone cell has its own specific function.

Osteocyte
(main bone tissue)

Osteoblast
(forms bone matrix)

Osteogenic cell
(stem cell)

Osteoclast
(reabsorbs bone)

Osteoblasts are responsible for forming new bone and are found in the growing portions of bone. Osteoblasts synthesize and secrete the collagen matrix. As the matrix surrounding the osteoblast calcifies, the trapped osteoblast changes in structure and becomes an osteocyte, the most common type of bone cell. Each osteocyte is located in a small cavity in the bone tissue called a lacuna. The cells responsible for breakdown and reabsorption are the osteoclasts, which originate from types of white blood cells. Osteoclasts break down old bone while osteoblasts form new bone. New bone tissue is constantly formed as old or damaged bone is dissolved for repair or to release calcium.

Osteoblasts and osteocytes are replaced by osteogenic cells, stem cells in the bone marrow that differentiate and develop into the specialized bone cells. Haematopoietic cells are also found in bone marrow; their function is to produce red blood cells, white blood cells and platelets.

BONE TYPES

Bones vary in their shape and function. They may be long, like the femur and the other bones of the arms and legs, or shorter, like the bones of the fingers and toes. The function of the long bones is to act as levers with the joints as fulcrums, pulled into position by muscle contractions. Short bones, like those in the wrists and ankles, are cuboid in shape. They allow some movement but mostly provide support and stability. Flat bones, which may often in fact be curved, include the shoulder blades, the breastbone and the bones of the skull. They provide protection for organs such as the heart and lungs and also act as attachment points for muscles.

Irregular bones, as the name suggests, do not have an easily described shape. The most prominent of these are the vertebrae of the spine which protect the spinal cord. Sesamoid bones, small round bones which are so named because their shape resembles that of a sesame seed, are not connected by joints to other bones. They are found embedded in tendons or muscles which they protect and support, such as the two pea-shaped bones located beneath the big toe joint in the ball of the foot. The largest of the sesamoid bones is the patella, or kneecap. Finally, sutural, or Wormian, bones are small additional bones sometimes found between the bones of the skull. The sutural and smaller sesamoid bones are not counted among the 206 principal bones of the body.

Long bones are divided into two main regions called the diaphysis and the epiphysis. The diaphysis is the hollow, tubular shaft that runs between the ends of the bone. In adults, the cavity inside the bone shaft is filled with yellow bone marrow, which is used for fat storage. The outer wall of the diaphysis is composed of dense and hard, compact bone tissue. The wider section at each end of the bone, called the epiphysis (plural: epiphyses), is filled with spongy bone. In some long bones the spaces between the spongy bone are filled by red bone marrow.

The epiphyseal plate, located where the epiphysis meets the diaphysis, is the site at which bone lengthening takes place through childhood and adolescence. The last bone to stop growing is the clavicle (collarbone) at around age 25. Although bones generally stop growing in length after adolescence, bone density will change continuously over the course of a lifetime as the bone tissue responds to stress from increased activity or changes in weight.

A layer of bone cells called the endosteum line the bone around the central cavity. These

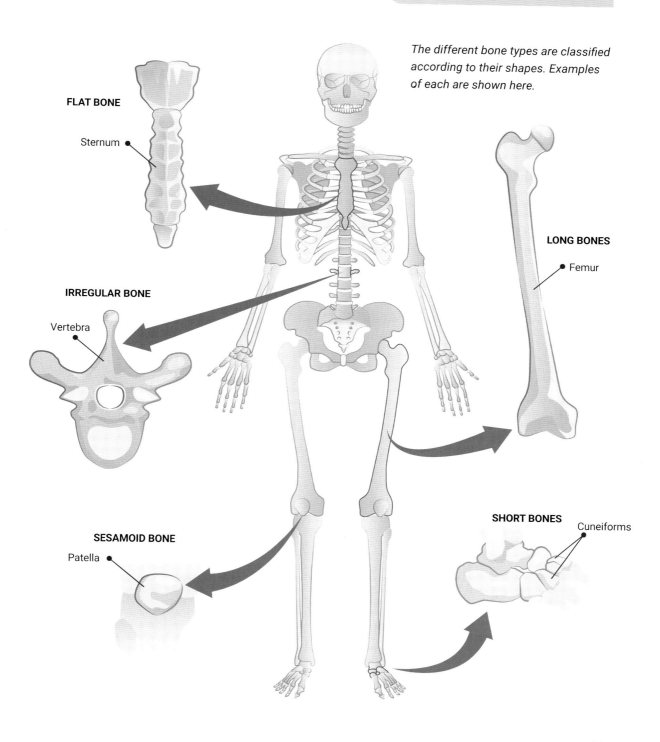

The different bone types are classified according to their shapes. Examples of each are shown here.

FLAT BONE

Sternum

IRREGULAR BONE

Vertebra

SESAMOID BONE

Patella

LONG BONES

Femur

SHORT BONES

Cuneiforms

59

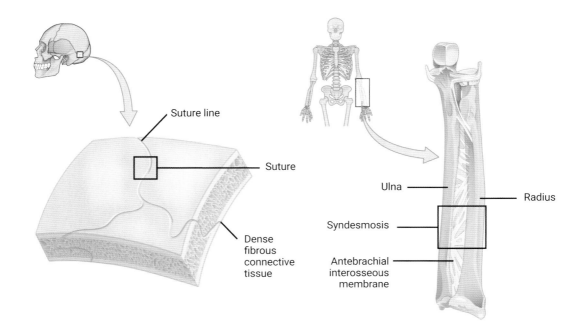

Suture line

Suture

Dense fibrous connective tissue

Ulna

Radius

Syndesmosis

Antebrachial interosseous membrane

cells are responsible for growth and repair of the bone throughout life. Another layer of cells on the outside of the bone, part of a structure called the periosteum, also take part in growth and repair. Tendons and ligaments attach to bones at the periosteum which forms a tough, thin, outer membrane that covers the entire outer surface of the bone except where the epiphyses of the bone meet other bones to form joints. Here, the epiphyses are covered with a thin layer of cartilage that reduces friction and acts as a shock absorber. Beneath the covering of the periosteum run nerves, along with blood vessels and lymphatic vessels that nourish the bone.

Flat bones, like those of the skull, are formed from a layer of spongy bone sandwiched on either side by a layer of compact bone. Even if the outer layer of the bone fractures, the intact inner layer of the bone will still provide protection for the internal organs.

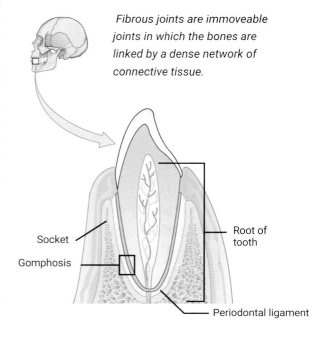

Fibrous joints are immoveable joints in which the bones are linked by a dense network of connective tissue.

Socket

Gomphosis

Root of tooth

Periodontal ligament

JOINTS

Bones link with other bones at joints. Apart from the hyoid bone in the neck, each of the 206 bones in the adult human body is connected to at least one other bone by a joint. Many joints allow the adjoining bones to move smoothly against each other, while others, joined firmly to each other by connective tissue or cartilage, allow little or no movement and provide stability.

Most of the joints of the skull, for example, are held rigidly together by fibrous connective tissue, which is important as the skull's purpose is to protect the brain. Similarly, the tibia and fibula of the leg are tightly joined, giving stability when standing. The joints of the vertebral column are linked by cartilage which allows for small movements between vertebrae that are just enough to permit the body to twist and bend.

There are three types of fibrous joint. The narrow fibrous joint found between most bones of the skull is called a suture. A joint such as that found between the long bones of the leg and arm, where the bones are separated but linked by a wide sheet of connective tissue, or a strap of fibrous connective tissue called a ligament, is called a syndesmosis. The narrow fibrous joint between the roots of a tooth and the bony socket in the jaw into which the tooth fits is called a gomphosis.

In a cartilaginous joint, as the name suggests, bones are united by cartilage, a tough type of connective tissue that has a degree of flexibility. Such joints are found in the ribcage and between the vertebrae of the spinal column. The pubic portions of the right and left hip bones of the pelvis are joined together by a cartilage pad. During pregnancy, increased levels of the hormone relaxin result in increased mobility of this joint allowing for expansion of the pelvic cavity during childbirth.

Synovial joints

The most common type of joint in the body is the synovial joint. The bones are not directly linked to each other but are connected by ligaments, which anchor the bones together, strengthening and supporting the joint but limiting the range of movement. The joint is enclosed within a cavity filled with lubricating synovial fluid, allowing the bones to move smoothly without rubbing against each other. The bones are covered with smooth cartilage where they meet, further reducing any friction between them. The joints between the bones of the appendicular skeleton are generally of this moveable type. Additional support at many synovial joints is provided by muscles and tendons, the dense connective tissue attaching muscle to bone, acting across the joint.

Below: Cartilaginous joints, as the name suggests, are joined by cartilage. They allow limited movement. Most of the vertebrae of the spine are linked by cartilaginous joints as are the bones of the pelvis at the pubic symphysis.

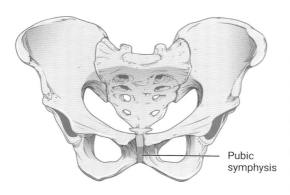

Pubic symphysis

A bursa (plural: bursae), a thin connective tissue sac filled with lubricating liquid, is usually found near joints and reduces friction by preventing skin, ligaments, muscles or tendons from rubbing against each other. A subcutaneous bursa is located between the skin and an underlying bone; a submuscular bursa is found between a muscle and a bone, or between adjacent muscles; a subtendinous bursa lies between a tendon and a bone. Bursae sometimes become inflamed, resulting in the painful condition of bursitis.

Synovial joint types

A pivot joint is one in which the rounded portion of one bone rotates around a single axis within a ring formed partially by a joint with another bone and partially by a ligament. The joint between the atlas and axis vertebrae in the neck which allows you to turn your head to follow the flight of the ball in a tennis match is a pivot joint.

In a hinge joint, the convex end of one bone links with the concave end of the adjoining bone, allowing bending and straightening motions along a single axis. The knee, elbow, and the joints in the fingers and toes are all examples of hinge joints.

A condyloid, or ellipsoid, joint is one in which a shallow depression at the end of one bone links with a rounded structure in another. Condyloid joints allow for movements in two planes. The knuckle joints of the hand are condyloid joints allowing the fingers to move up and down or to spread apart.

In a saddle joint, as the name suggests, the bones fit together like a rider sitting on a saddle. The first carpometacarpal joint, at the base of the thumb, allows the thumb to move both within the same plane as the palm of the hand, or to jut out perpendicular to the palm, giving humans their characteristic 'opposable' thumbs. Without the condyloid and saddle joints in the hand we would not have the manipulative skills essential to building human civilization.

A plane, or gliding, joint is one in which the surfaces of the bones are flat or slightly curved and can slide against each other. Motion at this type of joint can be multidirectional but is usually restricted by surrounding ligaments. Depending upon the specific joint of the body, a plane joint may allow movement in a single plane or in multiple planes. The carpal bones of the wrist and tarsal bones of the foot are linked by plane joints.

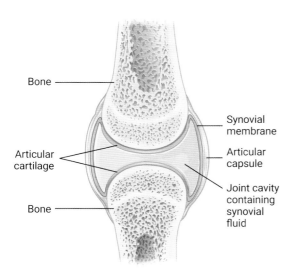

Bone

Articular cartilage

Bone

Synovial membrane

Articular capsule

Joint cavity containing synovial fluid

Synovial joints have the greatest degree of movement. A fluid-filled synovial cavity between the bones is enclosed by a membrane, providing extra cushioning to the ends of the bones which are held together by ligaments.

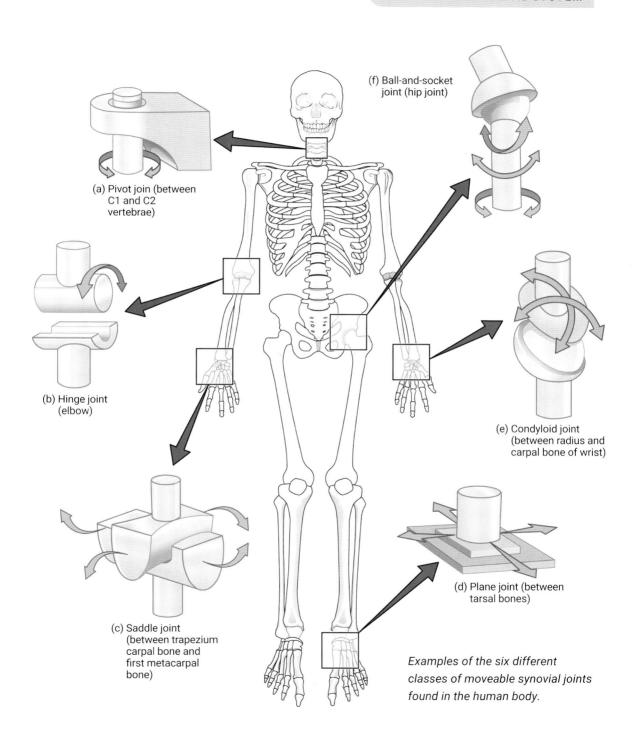

(f) Ball-and-socket joint (hip joint)

(a) Pivot join (between C1 and C2 vertebrae)

(b) Hinge joint (elbow)

(c) Saddle joint (between trapezium carpal bone and first metacarpal bone)

(d) Plane joint (between tarsal bones)

(e) Condyloid joint (between radius and carpal bone of wrist)

Examples of the six different classes of moveable synovial joints found in the human body.

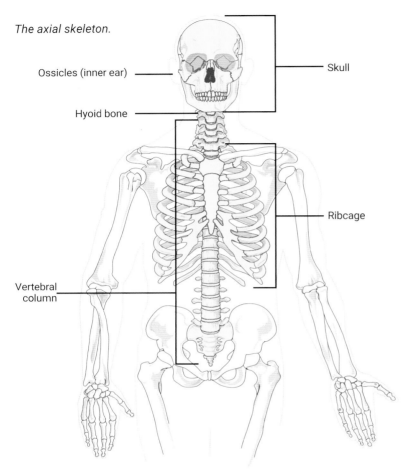

The axial skeleton.

Ossicles (inner ear)

Skull

Hyoid bone

Ribcage

Vertebral
column

spinal cord, heart and lungs. It provides the attachment site for the muscles that move the head, neck and back, as well as the muscles across the shoulder and hip joints that move the arms and legs. It includes all of the bones of the head, neck, chest and back. In adults it comprises the 22 bones of the skull, the 24 bones of the vertebral column, plus the fused vertebrae of the sacrum and coccyx, plus the 12 pairs of ribs and the sternum, or breastbone, that form the thoracic, or rib, cage. An additional seven bones are associated with the head, including the hyoid bone in the upper neck and the ear ossicles (three small bones found in each middle ear).

The Skull

The skull supports the face and protects the brain. In adults, it consists of 22 individual bones, 21 of which are fused into a single immobile unit. The only moveable bone in the skull is the mandible or lower jaw.

The facial bones provide support for the eyes, teeth and structures of the face with openings to allow for eating and breathing. The bony socket that houses the eyeball, along with the muscles that move it and open the upper eyelid, is called the orbit. Small openings above and below the orbit permit the passage of sensory nerves to the forehead and to the face below the orbit.

The ball-and-socket joint has the greatest range of motion. The rounded head of one bone is located within a socket in an adjacent bone. The only examples of such a joint in the human body are the hip and shoulder joints. The femur and the humerus can move both back and forth and from side to side and also have a degree of rotation around their long axis.

THE AXIAL SKELETON

As the name suggests, the axial skeleton forms the central axis of the body, protecting the brain,

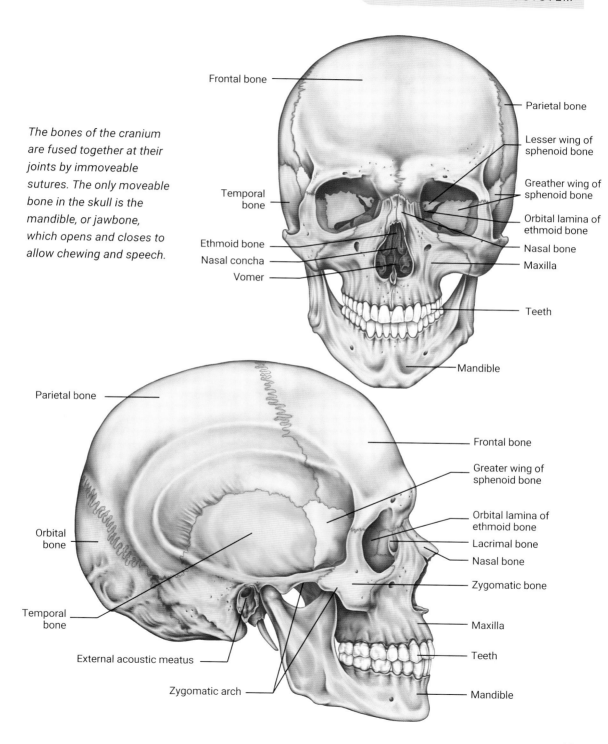

The bones of the cranium are fused together at their joints by immoveable sutures. The only moveable bone in the skull is the mandible, or jawbone, which opens and closes to allow chewing and speech.

Frontal bone

Parietal bone

Lesser wing of sphenoid bone

Greather wing of sphenoid bone

Orbital lamina of ethmoid bone

Nasal bone

Maxilla

Teeth

Mandible

Temporal bone

Ethmoid bone

Nasal concha

Vomer

Parietal bone

Frontal bone

Greater wing of sphenoid bone

Orbital lamina of ethmoid bone

Lacrimal bone

Nasal bone

Zygomatic bone

Maxilla

Teeth

Mandible

Orbital bone

Temporal bone

External acoustic meatus

Zygomatic arch

65

The nasal cavity is divided in half by the nasal septum, formed by the ethmoid bone and the vomer bone. Seen from the front of the skull, two bony plates can be seen projecting from each side wall of the nasal cavity.

A bridge of bone called the zygomatic arch, or cheekbone, separates the cranium and the upper and lower jaws and spans from the cheek to just above the ear canal. It is formed from the junction of the zygomatic bone and the temporal bone. The masseter muscle, which pulls the mandible upwards during biting and chewing, is anchored to the zygomatic arch.

The cranium houses and protects the brain. It consists of eight bones, including the paired parietal and temporal bones, along with the unpaired frontal, occipital, sphenoid and ethmoid bones. The interior space occupied by the brain is called the cranial cavity. The bones forming the top and sides of the cranium are usually referred to as the flat bones of the skull. The base of the skull, or cranial floor, has a number of openings to allow the passage of the spinal cord, cranial nerves and blood vessels.

The left and right parietal bones join together at the top of the skull at the sagittal suture. Each parietal bone joins the frontal bone at the coronal suture; the temporal bone, which forms the lower side of the skull, at the squamous suture; and the occipital bone at the lambdoid suture. The mastoid process, a muscle attachment site that projects from the temporal bone, can be felt just behind the earlobe on the side of the head. Small cavities within the temporal bone house the structures of the middle and inner ears.

The frontal bone forms the forehead. A slight depression, called the glabella, is located between the eyebrows. The frontal bone also forms the upper part of the orbit, thickening to form the rounded brow ridges, located just behind the eyebrows. The frontal bone extends back to form both the roof of the orbit below and the floor of the front of the cranial cavity above.

The occipital bone is a single bone forming the back of the skull. To either side of a bump called the external occipital protuberance is the highest point at which the muscles of the neck attach to the skull, with the scalp covering the skull above this. The occipital bone has a large opening called the foramen magnum, which allows the spinal cord to enter the skull. On either side of the foramen magnum the oval-shaped occipital condyles form joints with the first cervical vertebra that allow you to nod your head in agreement.

The sphenoid bone is like a keystone, joining with almost every other bone in the skull. It forms much of the base of the central skull and extends to form part of its sides. A region of the sphenoid bone at the base of the brain called the sella turcica ('Turkish saddle') houses the pea-sized pituitary gland, one of the major organs of the hormone-producing endocrine system.

The Vertebral Column

The vertebral, or spinal, column protects the spinal cord. It also provides structural support for the body, supporting around half of its

SUTURES

A suture is an immobile joint linking adjacent bones of the skull. A dense, fibrous connective tissue fills the gap between the bones, locking them together. The sutures follow irregular, twisting paths rather than straight lines making a tight, strong joint.

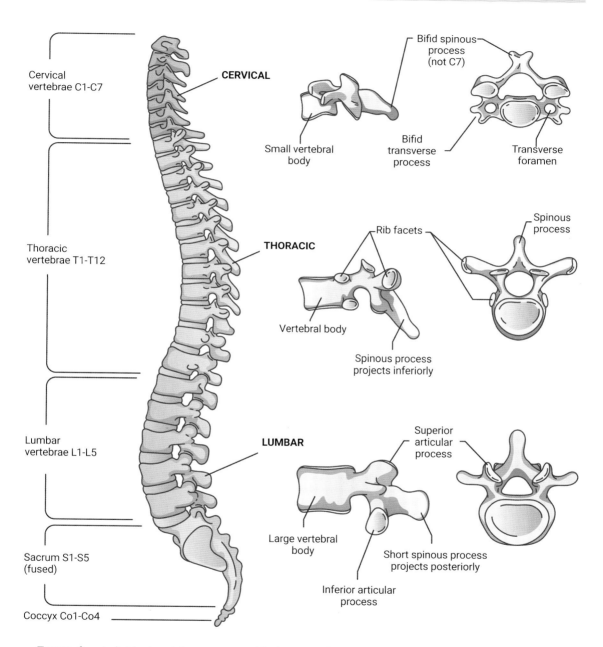

Cervical vertebrae C1-C7

Thoracic vertebrae T1-T12

Lumbar vertebrae L1-L5

Sacrum S1-S5 (fused)

Coccyx Co1-Co4

CERVICAL

Bifid spinous process (not C7)

Small vertebral body

Bifid transverse process

Transverse foramen

THORACIC

Rib facets

Spinous process

Vertebral body

Spinous process projects inferiorly

LUMBAR

Superior articular process

Spinous process

Large vertebral body

Inferior articular process

Short spinous process projects posteriorly

Twenty-four individual vertebrae separated by intervertebral discs of cartilage make up the vertebral column. Nine additional vertebrae are fused together at the base of the spine forming the sacrum and coccyx. The vertebrae are similar in structure but change in size and thickness reflecting their greater weight-bearing role further down the column. In profile, the vertebral column forms an S-shaped curve.

weight. The spine is flexible, both supporting the head, neck and body and allowing them to move. It is made up of a sequence of vertebrae (singular: vertebra), each separated and joined by a cartilaginous intervertebral disc. The spinal cord passes through an opening in each vertebra called the vertebral foramen. Together, these openings align to form the spinal canal.

At birth, the vertebral column has 33 vertebrae, but this eventually reduces to 24 as some fuse together to form the sacrum and coccyx, a process that starts around age 20 and completes in middle age. The vertebral column is subdivided into five regions. In the neck, there are seven cervical vertebrae, below these are the 12 thoracic vertebrae, and the five lumbar vertebrae in the lower back. The sacrum is formed by the fusion of the five sacral vertebrae, and the coccyx, or tailbone, from the fusion of the small coccygeal vertebrae.

Vertebrae vary in size and shape, but all have a similar structure. Cervical vertebrae are smaller than lumbar vertebrae because they support a smaller proportion of body weight. The body of the vertebra, at the front, is the weight-supporting part and gradually increases in size and thickness going down the vertebral column. A typical cervical vertebra has a small body, as it carries the least amount of body weight, and has openings for the arteries that supply the brain to pass through. The first cervical vertebra, called the atlas, supports the skull. The second cervical vertebra, called the axis, serves as the axis of rotation when turning the head to left or right.

The thoracic vertebrae have larger bodies than the cervical vertebrae. They have sites called facets, which are the points at which a rib is attached. Lumbar vertebrae have the largest and thickest vertebral bodies, reflecting the fact that they carry the greatest amount of body weight.

The vertebral arch to the rear of each vertebra serves as an important site for muscle attachment. The junctions between adjoining vertebrae form joints that play a major role in determining the type and range of motion available in each region of the spine.

The vertebrae of the sacrum and coccyx become fused together into single bones. The triangular-shaped sacrum, formed by the fusion of the five sacral vertebrae, is thick and wide at the top where it is weight bearing and then tapers down. The coccyx, or tailbone, is formed from the fusion of usually four very small coccygeal vertebrae. It links with the lower tip of the sacrum forming a slightly moveable joint. It is not weight bearing in the standing position but may take some body weight when sitting.

The vertebral column has four curvatures along its length which give it more strength, flexibility and ability to absorb shock than it would have if it were simply straight. The curvatures become more pronounced when carrying a heavy load and spring back when the load is removed.

The curvatures of the vertebral column are classified as either primary or secondary curvatures. Primary curvatures are retained from the original foetal curvature, while secondary curvatures develop after birth. During foetal development, the entire vertebral column has a single curvature. This is retained in the adult as the thoracic curve, which involves the thoracic vertebrae, and the sacrococcygeal curve, formed by the sacrum and coccyx. The cervical curve of the neck region forms as the infant develops the ability to hold their head upright when sitting. The lumbar curve of the lower back develops as the child begins to stand and then to walk.

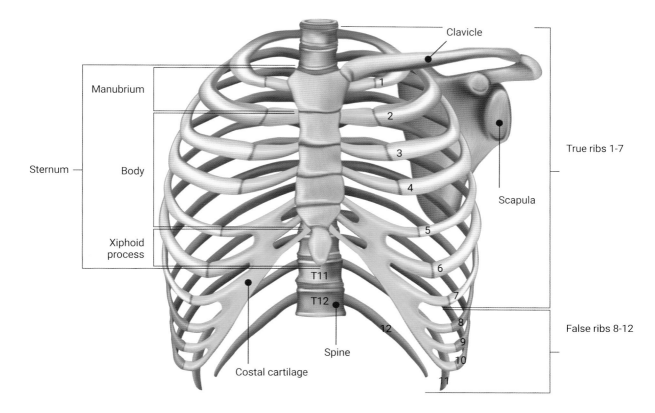

The bones of the thoracic cage protect the organs in the chest. The true ribs are attached to both the vertebral column and the sternum by cartilage. False ribs are attached to the vertebral column and the ribs above them, but not to the sternum. Floating ribs (ribs 11 and 12) end within the muscles of the abdominal wall.

In adults, the lumbar curve is generally deeper in females.

Adjoining vertebrae are anchored to each other by an intervertebral disc of cartilage which provides shock-absorbing padding and is also flexible enough to allow for limited movement between the vertebrae. Approximately 25 per cent of an adult's length from the top of the pelvis to the base of the skull is accounted for by the intervertebral discs, which are thinnest in the cervical region and thickest in the lumbar region. The tough, fibrous outer layer of the disc houses a softer, more gel-like material with a high water content that resists compression and is therefore important for weight bearing. The water content decreases with age causing the disc to become thinner, which decreases body height and reduces flexibility and range of motion. If the tough, outer layer is weakened due to injury or age, the pressure exerted on the disc when lifting a heavy object can cause the interior gel to protrude, resulting in the condition known as a 'slipped' disc.

Thoracic Cage

The thoracic, or rib, cage, formed from 12 pairs of ribs and the sternum, encloses and protects the heart and lungs.

The sternum, or breastbone, anchors the ribs at the front of the chest. It consists of three parts: the manubrium, the body, and the xiphoid process. The manubrium is the wider, upper part of the sternum and has a shallow, U-shaped border called the jugular notch at the base of the neck between the left and right clavicles, or collarbones. The first pair of ribs are attached to the manubrium but hidden behind the clavicle.

The body of the sternum is its elongated, central portion. The junction between the manubrium and the body forms a slight bend, called the sternal angle. The second pair of ribs, the highest pair that can be located by touch, are attached at the sternal angle. Pairs three to seven are attached to the sternal body. The lower end of the sternum is the xiphoid process, a small structure that is cartilaginous in early life, but becomes ossified around middle age.

The 12 pairs of ribs are curved, flattened bones attached to the thoracic vertebrae at the back, and in most cases to the sternum in front via a costal cartilage, which can extend for several centimetres. The ribs are classified into three groups based on how they relate to the sternum. Ribs 1 to 7, in which the costal cartilage attaches directly to the sternum, are classified as true ribs. In ribs 8–10, called false ribs, the costal cartilages do not attach directly to the sternum. For ribs 8–10, the costal cartilages are attached to the cartilage of the next higher rib, i.e. the cartilage of rib 10 attaches to the cartilage of rib 9 and rib 9 attaches to rib 8. Ribs 11 and 12 are also known as floating ribs. These short ribs do not attach to the sternum at all, but instead, their costal cartilages end within the musculature of the abdominal wall.

THE APPENDICULAR SKELETON

Attached to the central axial skeleton are the 126 bones that together make up the appendicular skeleton. These bones are divided into two groups: the girdle bones that attach the limbs to the axial skeleton and the bones that are located within the limbs themselves.

The Pectoral Girdle

The upper limbs are attached to the axial skeleton by the left and right pectoral, or shoulder, girdles, which each consist of two bones, the scapula, or shoulder blade, and clavicle. The pectoral girdles are not joined to each other and can operate independently. The S-shaped clavicle is located in front of the shoulder and is attached to the sternum, which is part of the axial skeleton, and to the scapula just above the shoulder joint. The clavicle of each pectoral girdle is anchored to the axial skeleton by a single, highly mobile joint that allows the shoulder and upper limb a great range of movement. Both clavicle and scapula are important attachment sites for muscles that move the shoulder and arm.

Clavicle

The clavicle is unusual in that it is the only long bone in the body that lies in a horizontal position. In men, the clavicle is heavier and longer, with a greater curvature and rougher surfaces where muscles attach. In women, the clavicle tends to be shorter, thinner and less curved. Anchored by muscles from above, it supports the scapula, holding the shoulder joint above and to the side of the trunk, which gives the upper limb the greatest freedom of movement. The clavicle also transmits

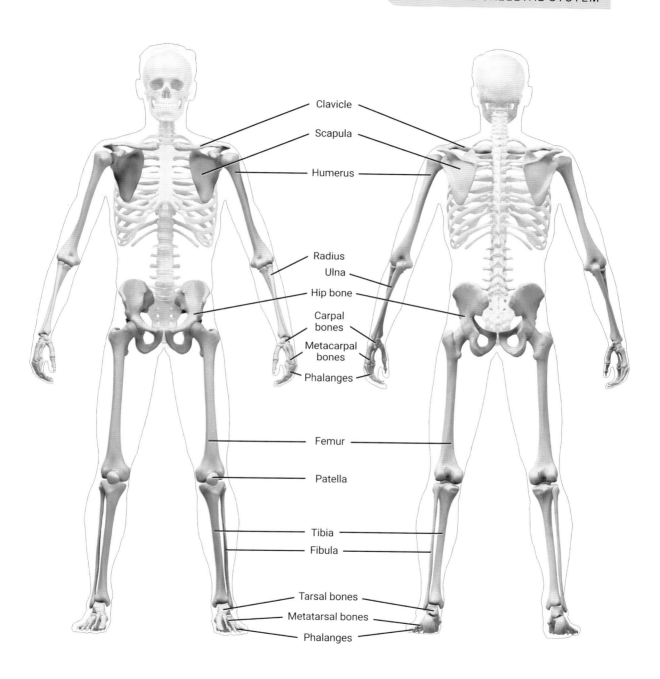

Clavicle

Scapula

Humerus

Radius

Ulna

Hip bone

Carpal
bones

Metacarpal
bones

Phalanges

Femur

Patella

Tibia

Fibula

Tarsal bones

Metatarsal bones

Phalanges

The 126 bones of the appendicular skeleton include all the bones of the limbs (arms, legs, hands and feet) along with the bones of the shoulder girdle and pelvic girdle.

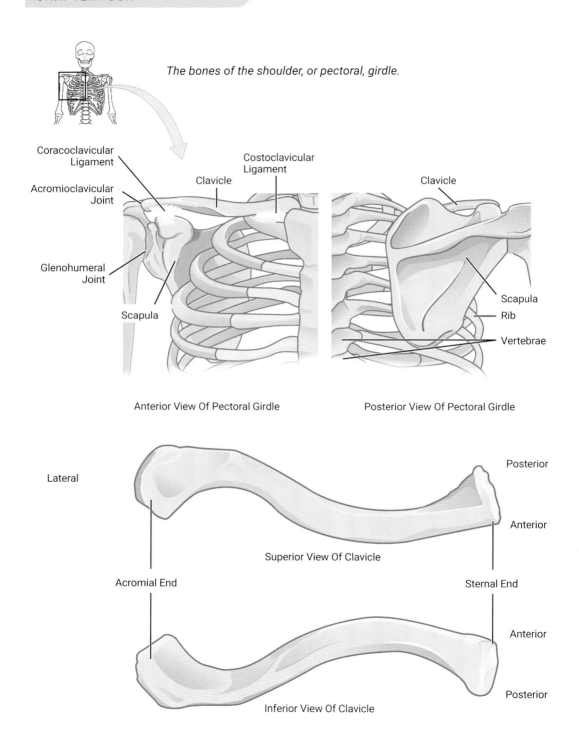

The bones of the shoulder, or pectoral, girdle.

Coracoclavicular Ligament

Costoclavicular Ligament

Clavicle

Acromioclavicular Joint

Clavicle

Glenohumeral Joint

Scapula

Scapula

Rib

Vertebrae

Anterior View Of Pectoral Girdle

Posterior View Of Pectoral Girdle

Lateral

Posterior

Anterior

Superior View Of Clavicle

Acromial End

Sternal End

Anterior

Inferior View Of Clavicle

Posterior

forces acting on the upper limb to the sternum and axial skeleton and protects the underlying nerves and blood vessels passing between the trunk of the body and the upper limb. The sternoclavicular joint, linking the clavicle and the sternum, allows the clavicle and scapula to move up, down, forward and back. It is strong and rarely dislocated, so excessive force, which might result from a fall on to an outstretched arm or from a strong blow, results in the breaking of the clavicle. The clavicle is the most commonly fractured bone in the body.

Scapula

The scapula is one of a pair of flat and roughly triangular bones located in the shoulders on the upper back of the trunk over ribs two to seven, but not attached to them. It serves as an attachment point for the powerful muscles involved in the movement of the shoulders and arms. The scapula is connected to the clavicle at the tip of the shoulder blade, called the acromion, by the acromioclavicular joint. The main support for the acromioclavicular joint comes from a very strong ligament called the coracoclavicular ligament. Located at the upper corner of the scapula is a shallow depression, called the glenoid cavity, where the scapula links with the humerus bone of the arm in the glenohumeral, or shoulder, joint.

Upper Limb Bones

The upper limb is divided into the arm, between the shoulder and elbow joints; the forearm, between the elbow and wrist joints; and the hand. There are 30 bones in each upper limb. The single bone of the upper arm is the humerus, and the ulna and radius are the paired bones of the forearm. At the base of the hand are the eight carpal bones, and the palm of the hand is formed by the five metacarpal bones. The fingers and thumb contain a total of 14 phalanx bones.

The humerus

The head of the humerus is large, round and smooth, linking with the glenoid cavity of the scapula to form the shoulder joint. Muscles that act across the shoulder joint are attached to the humerus. The lower end of the humerus joins the ulna and radius bones of the forearm to form the elbow joint. The pulley-shaped trochlea articulates with the ulna, while the knob-like capitulum links with the radius. Just above these areas are two small depressions into which the forearm bones fit when the elbow is fully bent. The humerus is sometimes referred to as the funny bone because the ulna nerve, which curves around the elbow end of the humerus, may cause an uncomfortable tingling sensation to run down the forearm if it is knocked suddenly.

Ulna and radius

The radius runs along the thumb side of the forearm with the ulna running parallel to it. The two bones are linked by a sheet of connective tissue. The C-shaped trochlear notch in the ulna joins with the trochlea of the humerus, while a small depression on the disc-shaped head of the radius joins with the capitulum of the humerus at the elbow joint. The lower end of the radius has a smooth surface that articulates with two carpal bones to form the radiocarpal, or wrist, joint.

Bones of the hand

Eight small carpal bones arranged in two rows form the wrist and base of the hand. The bones in the first row, beginning from the thumb side, are the scaphoid ('boat-shaped'), lunate ('moon-shaped'), triquetrum ('three-cornered') and pisiform ('pea-shaped') bones. The pisiform is the bony bump that can be felt at the base of your hand on the inner wrist. These four bones are linked together by ligaments in a single unit. The radius links to the scaphoid and lunate bones; the ulna does not directly link to any carpal bones. The second row of bones are the trapezium ('table'), trapezoid ('resembles a table'), capitate ('head-shaped') and hamate ('hooked') bones. These are also held together by ligaments and link with the metacarpal bones of the hand. Both rows of carpal bones link together as the midcarpal joint, which allows for all the movements of the hand at the wrist.

The carpal bones are grouped in a U-shape, with a strong ligament spanning the top of the U. Together, carpal bones and ligament form the carpal tunnel. The tendons of nine muscles of the forearm as well as an important nerve pass through the carpal tunnel into the hand. Carpal tunnel syndrome is a result of muscle tendon strain or wrist injury, causing inflammation and swelling resulting in pain or numbness, and muscle weakness in the hand.

Five metacarpal bones lie between the carpal bones in the wrist and the bones of the fingers and thumbs. Each metacarpal bone articulates at one end with one of the carpal bones at a carpometacarpal joint and at the other end with a phalanx bone of the thumb or one of the fingers at a metacarpophalangeal joint. The end joining the phalanx bone forms the knuckles of the hand. The first metacarpal bone, at the base of the thumb, is separated from the others, allowing it independent freedom of motion. The remaining metacarpals are united, forming the palm of the hand. The second and third metacarpals are immobile but the fourth and fifth have limited mobility that is important when the hand is used for gripping.

The 14 bones of the fingers and thumb are the phalanx bones (plural: phalanges). Digit number 1 (the thumb) has two phalanges, digits 2 to 5 (index finger to little finger) have three phalanges each, called the proximal, middle and distal phalanx bones. Phalanx bones link at the interphalangeal joints.

The Pelvic Girdle and Pelvis

The pelvic girdle, which acts as the attachment point for the lower limbs, supports the combined weight of the trunk and the head. It is formed by the two hip bones, large, curved bones formed by three separate bones – the ilium, ischium and pubis – that fuse together during the late teenage years. The ilium is the largest part of the hip bone and joins with the sacrum at the largely immobile sacroiliac joint. It provides attachment points for muscles in the thigh as well as the strong ligaments that support the sacroiliac joint. The ischium to the rear of the hip supports the body when sitting and provides attachment for the muscles at the rear of the thigh. The pubis at the front of the hip bone curves to join with the pubis of the opposite hip bone at the pubic symphysis.

The hip bones together with the sacrum and coccyx of the axial skeleton form the pelvis. In contrast to the highly mobile bones of the pectoral girdle, the bones of the pelvis are strongly united, forming a largely immobile structure that enables the weight of the body to be transferred from the vertebral column, through the pelvic girdle and hip joints, and into the weight-bearing lower limbs, giving strength and stability.

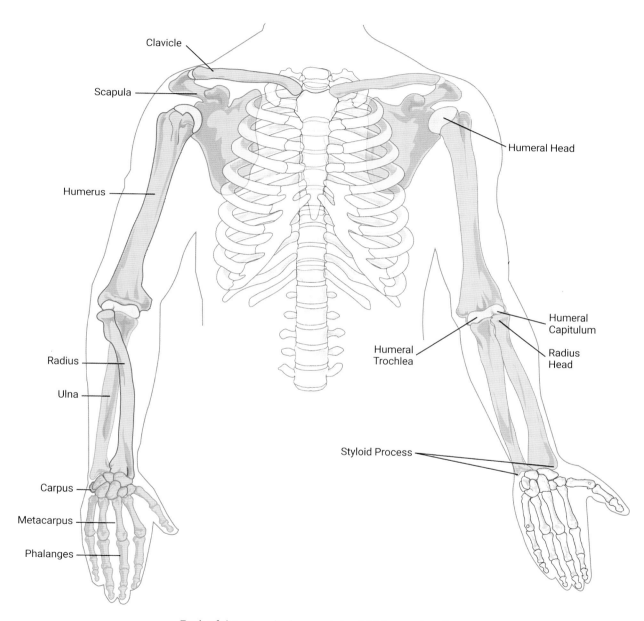

Clavicle

Scapula

Humeral Head

Humerus

Humeral Capitulum

Humeral Trochlea

Radius Head

Radius

Ulna

Styloid Process

Carpus

Metacarpus

Phalanges

Each of the upper limbs consists of 30 bones, 27 of which are in the wrist and hand.

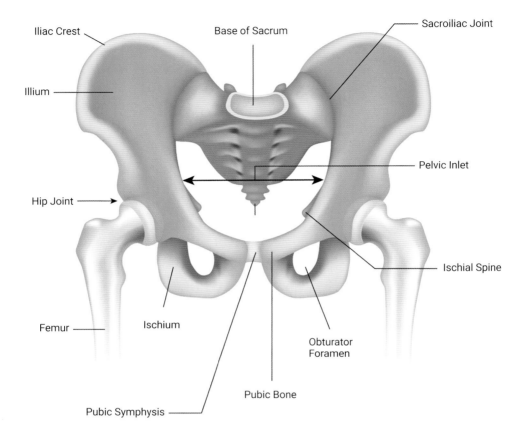

Iliac Crest

Base of Sacrum

Sacroiliac Joint

Illium

Pelvic Inlet

Hip Joint

Ischial Spine

Femur

Ischium

Obturator
Foramen

Pubic Bone

Pubic Symphysis

As well as providing attachment for the legs, the pelvic girdle contains and
protects the organs of the abdomen. Each half of the girdle is attached to the
vertebral column by ligaments and attached to each other by the pubic symphysis.
Adapted for childbirth, the adult female pelvis is wider than that of the male.

The pelvis also gives protection to the internal organs. The broad region covered by the fan-like upper hip bone, called the greater pelvis or greater pelvic cavity, is occupied by part of the small and large intestines. The narrow, rounded space of the lesser pelvis, or lesser pelvic cavity,

beneath encloses the bladder and other pelvic organs.

There are distinct differences between the adult female and male pelvis. In general, the bones of the male pelvis are thicker and heavier, adapted to support the generally heavier build

Thirty bones make up each lower limb, 26 of which are in the ankle and foot.

of the male. The female pelvis is adapted for childbirth and is wider than the male pelvis. The female sacrum is wider, shorter and less curved than that of the male, and the lesser pelvic cavity is also wider and shallower than the narrower, deeper lesser pelvis of males. Because of these obvious differences the hip bone is the one bone of the body that allows for accurate sex determination.

Lower Limb Bones

The lower limb, like the upper limb, is divided into three regions. The thigh is located between the hip joint and knee joint, the leg comes between the knee joint and the ankle joint, anatomically speaking, and finally comes the foot. The lower limb contains 30 bones: the femur, the single bone of the thigh; the patella, or kneecap; the tibia, the large weight-bearing bone of the leg and the thinner fibula; and the seven tarsal bones and five elongated metatarsal bones of the foot and 14 small phalanges in the toes.

Femur

The femur, or thigh bone, is the longest and strongest bone in the body, accounting for approximately

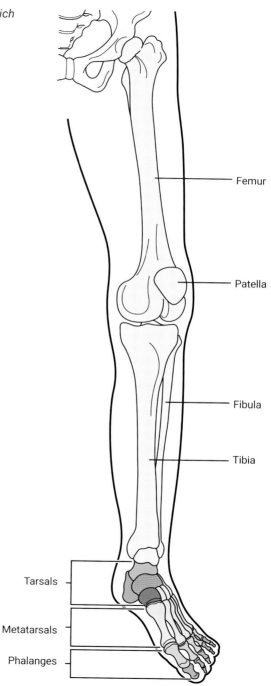

Femur

Patella

Fibula

Tibia

Tarsals

Metatarsals

Phalanges

one-quarter of an adult's total height. The rounded head of the femur fits into a cup-shaped cavity in the hip bone called the acetabulum to form the hip joint, a ball-and-socket type joint. The narrowed region below the head is the neck of the femur, a common area for fractures of the femur. A bony projection called the greater trochanter extending upwards above the base of the neck provides attachment and leverage for multiple muscles acting across the hip joint. The smaller lesser trochanter, just below the neck of the femur is the anchor for a single, powerful muscle.

Patella

The patella, or kneecap, is the largest of the sesamoid bones in the body. A sesamoid bone is one that is incorporated into the tendon of a muscle where that tendon crosses a joint and prevents damage to the muscle tendon due to rubbing against the bones during movements of the joint. The patella is located in the tendon of the quadriceps femoris muscle, the large muscle at the front of the thigh that passes across the front of the knee to attach to the tibia. As well as protecting the tendon from damage, the patella also lifts it away from the knee joint, increasing the leverage power of the quadriceps femoris muscle as it acts across the knee.

Tibia

The tibia, or shin bone, is paired with the smaller fibula. It is the second longest bone in the body after the femur and is the main weight-bearing bone of the leg. The upper end of the tibia links with the femur to form the knee joint. A sheet of dense connective tissue unites the tibia and fibula bones. At the lower end of the tibia the medial malleolus ('little hammer'), which forms the large bony bump found in the ankle region, links with the talus bone of the foot as part of the ankle joint.

Fibula

The fibula is a slender bone and does not bear weight. It is primarily employed for muscle attachments and links with the talus bone of the foot as part of the ankle joint.

Bones of the foot

The seven tarsal bones form the rear half of the foot. The uppermost bone is the talus, the square-shaped, upper surface of which links with the tibia and fibula to form the ankle joint. The talus also links with the calcaneus, or heel bone, the largest bone of the foot. Body weight is transferred from the tibia to the talus to the calcaneus, which rests on the ground.

The front half of the foot is formed by the five metatarsal bones, located between the tarsal bones and the phalanges of the toes. These join with the cuboid or cuneiform bones. The first metatarsal bone is shorter and thicker than the others and the second metatarsal is the longest. The base of the fifth metatarsal has a large expansion for muscle attachments. The heads of the metatarsal bones rest on the ground forming the ball of the foot. The 14 phalanx bones of the toes are arranged in a similar manner to the phalanges of the fingers. The toes are numbered one to five, starting with the hallux, or big toe. The big toe has two phalanx bones, the proximal and distal phalanges. The remaining toes all have proximal, middle and distal phalanges. The joint between phalanx bones is called an interphalangeal joint.

The interlocking bones of the foot provide both weight-bearing strength and flexibility.

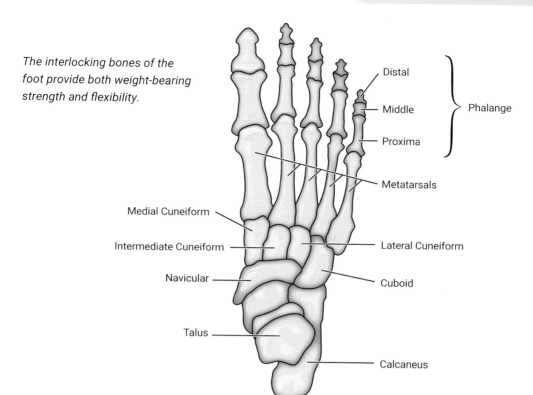

Distal
Middle
Proxima
Phalange
Metatarsals
Medial Cuneiform
Intermediate Cuneiform
Lateral Cuneiform
Navicular
Cuboid
Talus
Calcaneus

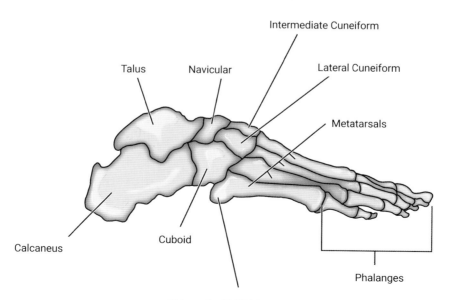

Intermediate Cuneiform
Talus
Navicular
Lateral Cuneiform
Metatarsals
Calcaneus
Cuboid
Phalanges
Tuberosity Of Fifth Metatarsal

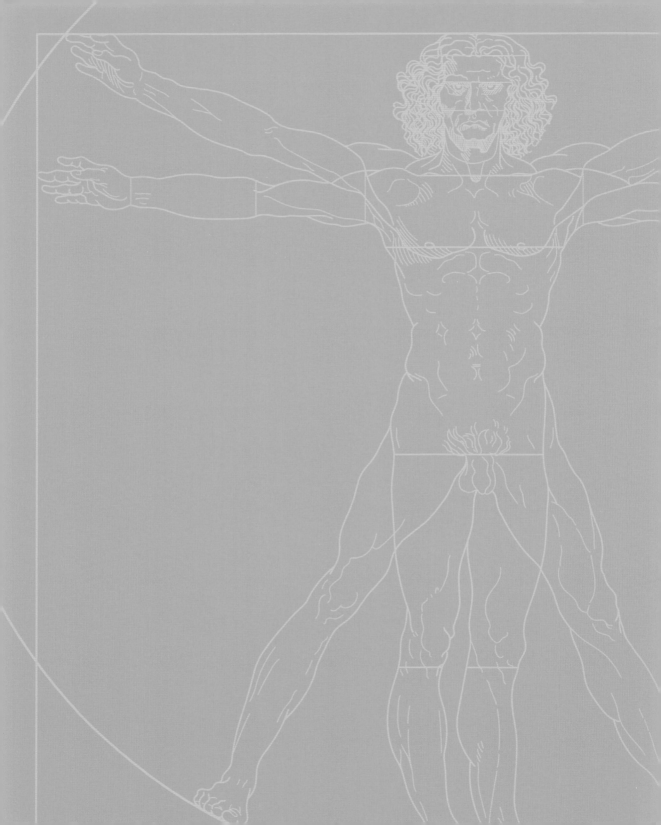

The Muscular System

The Muscular System

Muscle cells are specialized for contraction and thereby causing various parts of the body to move, whether it be for walking, lifting, breathing or transporting food through the digestive system. All of the muscles in the human body together make up the muscular system.

the body upright and balanced. Skeletal muscles also protect and support internal organs, acting as a shield against blows. Over 40 per cent of

SKELETAL MUSCLE

There are three types of muscle in the human body. The muscles with which we are most familiar are the skeletal muscles, so-called because most of them are attached to and move the skeleton. They are also sometimes known as voluntary muscles because they are, by and large, under conscious control. There are around 650 skeletal muscles, and these are the ones we use when we run, jump, chew our food or scratch our heads. Constant small adjustments of the skeletal muscles, that happen without us really having to think about it, maintain our posture, keeping

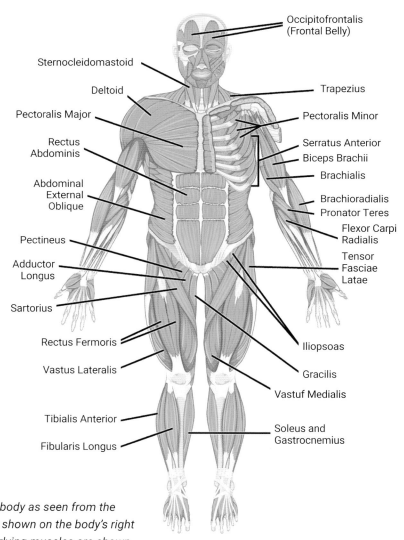

Occipitofrontalis (Frontal Belly)

Sternocleidomastoid

Deltoid

Trapezius

Pectoralis Major

Pectoralis Minor

Rectus Abdominis

Serratus Anterior

Biceps Brachii

Brachialis

Abdominal External Oblique

Brachioradialis

Pronator Teres

Flexor Carpi Radialis

Pectineus

Tensor Fasciae Latae

Adductor Longus

Sartorius

Rectus Fermoris

Iliopsoas

Vastus Lateralis

Gracilis

Vastuf Medialis

Tibialis Anterior

Soleus and Gastrocnemius

Fibularis Longus

The major muscles of the human body as seen from the front. The superficial muscles are shown on the body's right side, while on its left deeper underlying muscles are shown.

an average adult male's body weight consists of skeletal muscle; females are slightly less at around 35 per cent.

The muscular and skeletal systems are often considered together as the musculoskeletal system. Skeletal muscles are attached to the skeleton by tough connective tissues called tendons. Many skeletal muscles are attached to the ends of bones at a joint where they span the joint and connect the bones. When the muscles contract, they pull on the bones, causing them to move. The skeletal system can be thought of

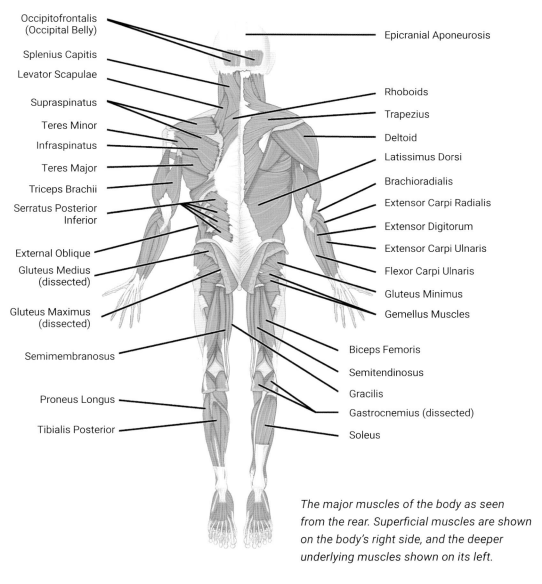

Occipitofrontalis (Occipital Belly)
Splenius Capitis
Levator Scapulae
Supraspinatus
Teres Minor
Infraspinatus
Teres Major
Triceps Brachii
Serratus Posterior Inferior
External Oblique
Gluteus Medius (dissected)
Gluteus Maximus (dissected)
Semimembranosus
Proneus Longus
Tibialis Posterior

Epicranial Aponeurosis
Rhoboids
Trapezius
Deltoid
Latissimus Dorsi
Brachioradialis
Extensor Carpi Radialis
Extensor Digitorum
Extensor Carpi Ulnaris
Flexor Carpi Ulnaris
Gluteus Minimus
Gemellus Muscles
Biceps Femoris
Semitendinosus
Gracilis
Gastrocnemius (dissected)
Soleus

The major muscles of the body as seen from the rear. Superficial muscles are shown on the body's right side, and the deeper underlying muscles shown on its left.

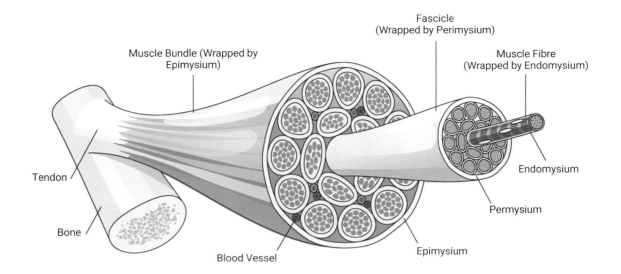

Muscle Bundle (Wrapped by Epimysium)

Fascicle (Wrapped by Perimysium)

Muscle Fibre (Wrapped by Endomysium)

Tendon

Bone

Blood Vessel

Epimysium

Permysium

Endomysium

The structure of a skeletal muscle is one of bundles within bundles. The long, string-like cells of muscle fibres are wrapped in connective tissue called endomysium; bundles of these fibres, each wrapped by perimysium, another connective tissue, form muscle fascicles and the fascicles are again bundled together to form the skeletal muscle, wrapped by epimysium, also a connective tissue.

as a system of levers that allow body movement, with the muscular system providing the force to move the levers.

The other two muscle types aren't under our conscious control but have vitally important roles to play. Cardiac muscle, found only in the heart, pumps blood through the circulatory system with strong, rhythmical and completely involuntary contractions. Smooth, or visceral, muscle is also concerned with various involuntary movements, such as those of the digestive system. Smooth muscle in the walls of arteries regulates blood pressure and blood flow through the circulatory system. Neither cardiac nor smooth muscle connect to bone.

A property common to all three types of muscle is their ability to contract and thus generate force which produces movement. As well as being able to shorten by contracting, smooth muscle tissue also has the ability to stretch before recoiling back to its original length.

In order to contract, skeletal muscles need stimulation from motor neurons. The point where a motor neuron attaches to a muscle is called a neuromuscular junction. If you decide to kick a ball, for instance, your brain sends electrical messages through motor neurons to the muscles in your leg, stimulating the muscle fibres to contract and causing your leg to kick.

Involuntary contractions of smooth and

HYPERTROPHY AND ATROPHY

The size of a muscle is the main determinant of its strength, the amount of force it can exert. Muscles can grow larger, or hypertrophy, through increased use. Running, or any other exercise that increases the heart rate, can increase the size and strength of cardiac muscle. Regular physical exercise can increase the size, and therefore the strength of skeletal muscles. Hormones can also play a role in muscle size; the increase in testosterone levels at male puberty leads to a significant increase in muscle mass.

Muscles can also shrink, or atrophy, from lack of use or as the result of a poor diet. People confined to bed through illness or injury lose muscle mass quickly. Muscle atrophy also occurs in old age, a condition known as sarcopenia, the exact cause of which is not known.

Neither hypertrophy nor atrophy involves a change in the number of muscle fibres. In hypertrophy the fibres become wider and with atrophy they become narrower, but in both instances the number of fibres stays the same.

The axons of a motor neuron (shown in yellow) send impulses in the form of chemical signals to muscle fibres, causing them to contract. The site where this takes place is called a neuromuscular junction.

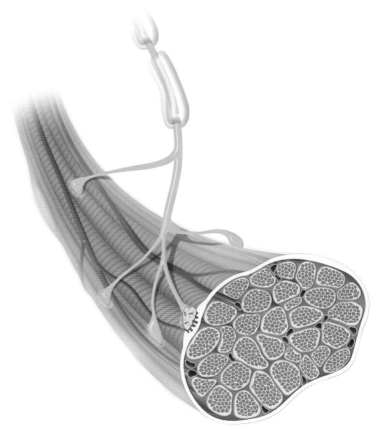

A muscle fibre is formed from bundles of myofibrils, which are in turn formed from bundles of thin filaments of the protein actin and thick filaments of the protein myosin. These filaments are arranged in repeating units called sarcomeres, the basic units of skeletal and cardiac muscle. When the muscle contracts the thick and thin filaments slide over each other, the myosin protein pulling on the thin actin filament.

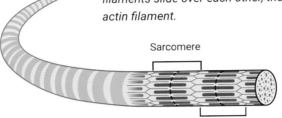

Sarcomere

Sarcomere (contractile unit of the myofibril)

Myofibril or **fibril** (complex organelle composed of bundles of myofilaments)

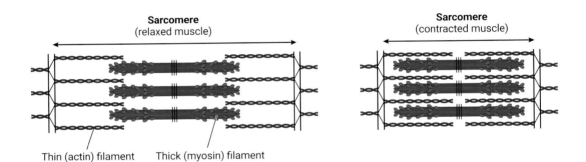

Sarcomere
(relaxed muscle)

Sarcomere
(contracted muscle)

Thin (actin) filament   Thick (myosin) filament

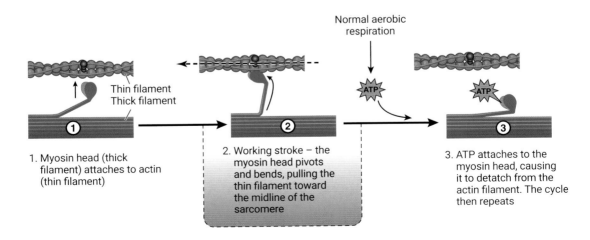

Normal aerobic respiration

Thin filament
Thick filament

ATP   ATP

1. Myosin head (thick filament) attaches to actin (thin filament)

2. Working stroke – the myosin head pivots and bends, pulling the thin filament toward the midline of the sarcomere

3. ATP attaches to the myosin head, causing it to detach from the actin filament. The cycle then repeats

cardiac muscles are also controlled by electrical impulses, but in the case of these muscles, these come from the autonomic nervous system in the case of smooth muscle or specialized pacemaker cells in the heart.

Skeletal and cardiac muscles are composed mainly of muscle cells, also called muscle fibres. A muscle fibre consists of a bundle of myofibrils, which are themselves bundles of protein filaments. These protein filaments consist of thin filaments of the protein actin, which are anchored to structures called Z discs, and thick filaments of the protein myosin. The protein filaments are arranged together within a myofibril in repeating units called sarcomeres, the basic functional units of skeletal and cardiac muscle. In both skeletal and cardiac muscle, the actin and myosin proteins are arranged in regular patterns, creating an alternating light and dark, striped pattern called striations. These striations are absent in smooth muscle as the contractile proteins are not arranged in a regular fashion. The mechanical process of contracting takes place when actin is pulled by myosin, using energy from ATP, so that the actin and myosin filaments slide over one another. The sliding filaments increase the tension in, and so shorten the length of,

the muscle cells. These muscle contractions are responsible for practically all of the movements that take place in the body, both inside and out.

Unlike other cell types, which have a single nucleus, skeletal muscle cells have multiple nuclei. Cardiac muscle cells have one centrally located nucleus, but the cells are physically and electrically connected to each other so that the entire heart contracts as one unit. Smooth muscle cells, called myocytes, also contain a single nucleus and can exist in electrically linked units contracting together as a single-unit or as multi-unit smooth muscle where the cells are not electrically linked.

Other tissues in the muscular system include the connective tissue that forms a protective and

The quadriceps tendon connects the powerful quadriceps muscles in the thigh to the top of the patella, or kneecap. The patellar ligament connects the patella to the shinbone. Working together, muscles, tendons and ligaments produce the movements of the knee joint that allow us to walk, run, kick and jump.

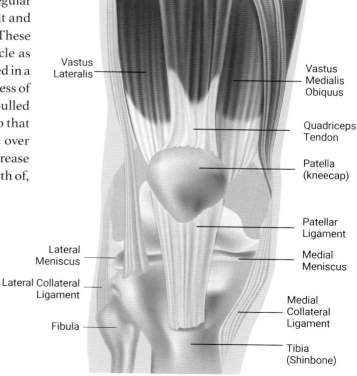

Vastus Lateralis
Vastus Medialis Obiquus
Quadriceps Tendon
Patella (kneecap)
Patellar Ligament
Medial Meniscus
Lateral Meniscus
Lateral Collateral Ligament
Medial Collateral Ligament
Fibula
Tibia (Shinbone)

FAST AND SLOW

Skeletal muscle fibres can be divided into two types: slow-twitch (or type I) and fast-twitch (or type II). Slow-twitch muscle fibres are well-supplied with capillaries and a large number of mitochondria as well as myoglobin, a protein that stores oxygen until it is needed. This means that slow-twitch fibres can sustain aerobic activity and can contract for long periods of time. These are the muscles used by endurance athletes such as marathon runners.

Fast-twitch muscle fibres have fewer capillaries and mitochondria and less myoglobin. They can contract rapidly and powerfully but can only sustain anaerobic (non-oxygen-using) activity for short periods. Fast-twitch fibres contribute more to muscle strength than slow-twitch fibres and come into play in activities requiring power over short periods such as sprinting.

The proportion of fast to slow fibres varies from muscle to muscle and from person to person. Generally, a person with more slow-twitch fibres is better suited for endurance activities whereas someone with more fast-twitch fibres will be better at activities requiring short bursts of power, such as weightlifting.

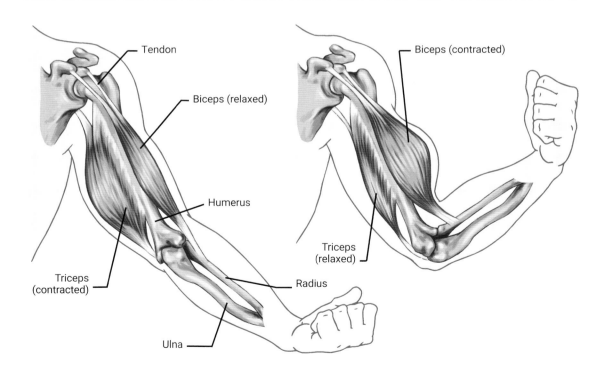

Rather than being arranged in regular rows like skeletal muscle, cardiac muscle has a highly branched structure that allows signals to propagate faster through the tissue and ensures that the cells contract together.

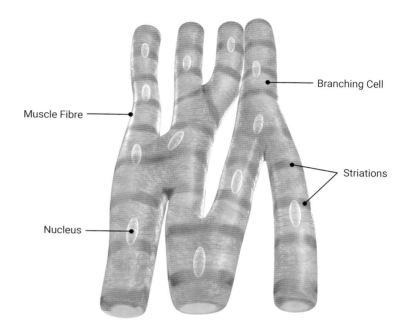

Branching Cell

Muscle Fibre

Striations

Nucleus

supporting sheath called epimysium around the muscle and the tendons formed from bundles of collagen fibres that attach muscles to bone. Each skeletal muscle consists of hundreds, sometimes thousands, of long, thin muscle fibres. These muscle fibres are bundled together in groups of ten to a hundred or more in units called muscle fascicles, which are surrounded by sheaths of connective tissue called perimysium. In turn, bundles of fascicles form individual skeletal muscles.

Skeletal muscles often work in pairs in opposition to each other, called antagonistic

muscle pairs, allowing certain bones to be moved back and forth. A familiar example of such a pair are the biceps and triceps muscles in the upper arm. Contracting the biceps at the front of the upper arm causes the arm to flex at the elbow and bend upwards. Contracting the triceps at the back of the arm straightens the arm out again.

CARDIAC MUSCLE

Although the heart is not the most powerful muscle in the human body, it has to work continuously over an entire lifetime. Cardiac muscle, or myocardium, is found only in the

Opposite: Because muscles can only pull and not push, muscles that move a joint usually work together in what is known as an antagonistic pair. As one muscle, called the agonist, contracts, the second muscle, called the antagonist, relaxes. A typical example of an antagonistic pair are the biceps muscle that contracts to raise the forearm (while the triceps relaxes), and the triceps muscle that contracts to straighten the arm again (while the biceps relaxes).

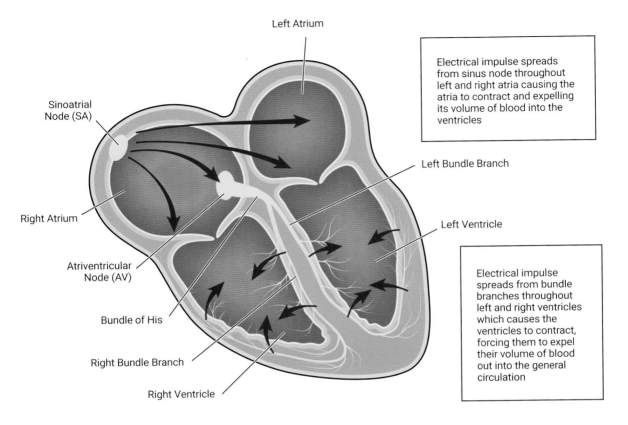

Left Atrium

Electrical impulse spreads from sinus node throughout left and right atria causing the atria to contract and expelling its volume of blood into the ventricles

Sinoatrial Node (SA)

Left Bundle Branch

Right Atrium

Left Ventricle

Atriventricular Node (AV)

Electrical impulse spreads from bundle branches throughout left and right ventricles which causes the ventricles to contract, forcing them to expel their volume of blood out into the general circulation

Bundle of His

Right Bundle Branch

Right Ventricle

The rhythmical beating of the heart, called sinus rhythm, is maintained by electrical signals generated by pacemaker cells located in the sinoatrial node. These signals travel rapidly, first to the right and left atria which contract together, then to the atrioventricular node, and from there via the bundle of His, to the right and left ventricles which contract together a split second after the contraction of the atria.

wall of the heart, enclosed within connective tissues, the endocardium on the inside and the pericardium on the outside of the heart. Cardiac muscle contains a great many mitochondria, which produce ATP for the energy required for the heart's continuous output. Like skeletal muscle, cardiac muscle is striated with filaments arranged in sarcomeres inside the muscle fibres, although the myofibrils in cardiac muscle branch at irregular angles rather than being arranged in parallel rows as they are in skeletal muscle.

Contractions of cardiac muscle are involuntary, controlled by electrical impulses from specialized cardiac muscle cells in a part of the heart called the sinoatrial node. The cells of cardiac muscle tissue are linked in interconnected networks that allow electrical impulses to propagate rapidly, ensuring that the contraction of the cells is co-ordinated and occurs virtually simultaneously.

SMOOTH MUSCLE

Contractions of smooth muscles are involuntary, controlled by the autonomic nervous system, hormones and neurotransmitters, which are not under conscious control. Unlike skeletal muscle, smooth muscle can sustain very long-term contractions. It can also stretch and still contract, which skeletal muscle cannot do. This is important in the functioning of organs such as the bladder and the uterus which are required to stretch considerably.

Smooth muscle is found throughout the body. For example, in the walls of organs of the gastrointestinal tract, such as the oesophagus, stomach and intestines, where it moves food through the tract by wave-like contractions called peristalsis. It is also found in the air passages of the respiratory tract, such as the bronchi, controlling their diameter and thus the volume of air that can pass through them. Smooth muscle in the walls of the uterus pushes the baby out into the birth canal at birth. In the walls of the urinary bladder, smooth muscle allows the bladder to expand to hold more urine, and then contract as the urine is expelled from the body. Flow through blood vessels and lymphatic vessels is also controlled by smooth muscles lining the walls of the vessels.

Smooth muscle cells do not have their filaments arranged in sarcomeres, so smooth tissue does not have the striated appearance of skeletal muscle, although they do contain myofibrils, with bundles of myosin and actin filaments which slide over each other resulting in contractions. The thin actin filaments are anchored to features called dense bodies which are analogous to the Z discs in skeletal muscle.

Intermediate filaments are networked through the cells of smooth muscle, anchored to structures called dense bodies. Thick filaments pulling on the dense bodies cause the cell to contract.

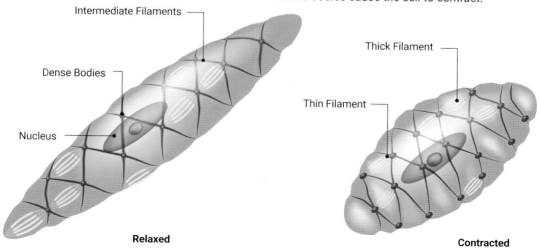

Intermediate Filaments

Dense Bodies

Nucleus

Thick Filament

Thin Filament

Relaxed

Contracted

The Nervous System

The Nervous System

The nervous system is responsible for co-ordinating the actions, both voluntary and involuntary, undertaken by the human body. The nervous system collects information from the internal and external environments using an array of sensory receptors. It sends this data to the brain, which assesses it and sends signals to muscles, organs or glands to bring about the appropriate response. These transmissions are carried through the medium of electrical signals travelling to and from different parts of the body through a complex system of specialized cells.

The nervous system can be split into two main divisions: the central nervous system (CNS), which includes the brain and spinal cord, and the peripheral nervous system (PNS), which connects the central nervous system to the rest of the body. The PNS is subdivided into the somatic and autonomic nervous systems. The somatic nervous system controls activities that are under voluntary control, such as lifting an apple

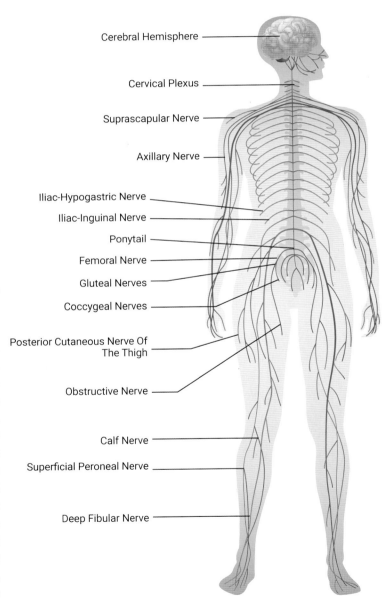

Cerebral Hemisphere

Cervical Plexus

Suprascapular Nerve

Axillary Nerve

Iliac-Hypogastric Nerve

Iliac-Inguinal Nerve

Ponytail

Femoral Nerve

Gluteal Nerves

Coccygeal Nerves

Posterior Cutaneous Nerve Of The Thigh

Obstructive Nerve

Calf Nerve

Superficial Peroneal Nerve

Deep Fibular Nerve

to your mouth to take a bite. The autonomic nervous system controls involuntary activities, such as the process of digesting the apple. The autonomic nervous system is further divided into the sympathetic nervous system, which controls the fight-or-flight response, the parasympathetic nervous system, which controls the routine functions of the body, and the enteric nervous

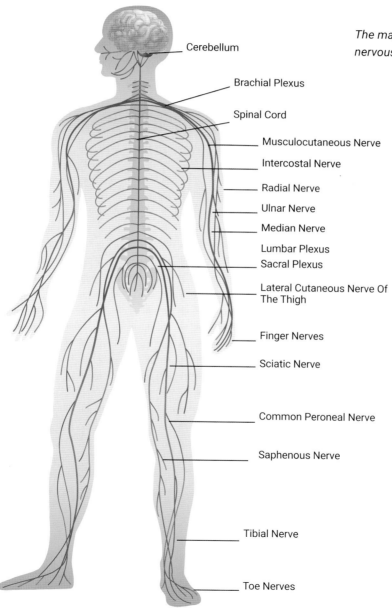

The major nerves of the human nervous system.

Cerebellum

Brachial Plexus

Spinal Cord

Musculocutaneous Nerve

Intercostal Nerve

Radial Nerve

Ulnar Nerve

Median Nerve

Lumbar Plexus

Sacral Plexus

Lateral Cutaneous Nerve Of The Thigh

Finger Nerves

Sciatic Nerve

Common Peroneal Nerve

Saphenous Nerve

Tibial Nerve

Toe Nerves

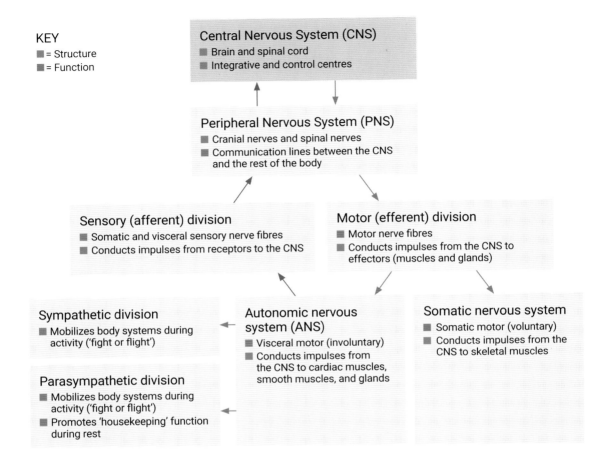

KEY
■ = Structure
■ = Function

Central Nervous System (CNS)
■ Brain and spinal cord
■ Integrative and control centres

Peripheral Nervous System (PNS)
■ Cranial nerves and spinal nerves
■ Communication lines between the CNS and the rest of the body

Sensory (afferent) division
■ Somatic and visceral sensory nerve fibres
■ Conducts impulses from receptors to the CNS

Motor (efferent) division
■ Motor nerve fibres
■ Conducts impulses from the CNS to effectors (muscles and glands)

Sympathetic division
■ Mobilizes body systems during activity ('fight or flight')

Parasympathetic division
■ Mobilizes body systems during activity ('fight or flight')
■ Promotes 'housekeeping' function during rest

Autonomic nervous system (ANS)
■ Visceral motor (involuntary)
■ Conducts impulses from the CNS to cardiac muscles, smooth muscles, and glands

Somatic nervous system
■ Somatic motor (voluntary)
■ Conducts impulses from the CNS to skeletal muscles

The nervous system divides into the central nervous system, which includes the brain and spinal cord, and the peripheral nervous system, further divided into the autonomic and somatic systems, extending throughout the rest of the body.

system, responsible for the activities of the digestive system.

NEURONS

The electrical signals sent through the nervous system are called nerve impulses, and are transmitted by cells called neurons, or nerve cells. A cell that receives nerve impulses from a neuron will be stimulated to perform a function, perhaps a muscle cell stimulated to contract, or a gland stimulated to release a hormone. The information transmitted by the nervous system

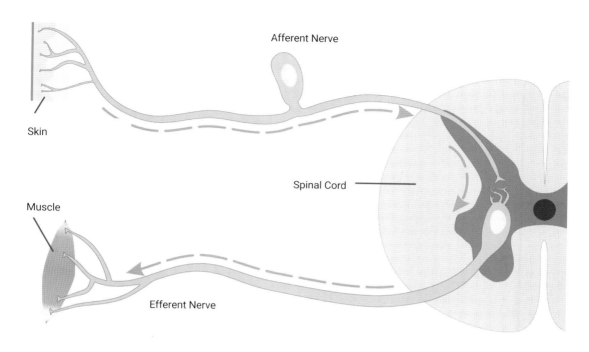

Afferent Nerve

Skin

Spinal Cord

Muscle

Efferent Nerve

Afferent, or sensory, nerves carry information from sensory receptors to the central nervous system, while efferent, or motor, nerves carry instructions back from the central nervous system to muscles and glands.

is transmitted very rapidly; the fastest impulses may travel through the body at over 100 metres (328 ft) per second.

Features of the neuron include a cell body, or soma, which houses most of the neuron's organelles such as its nucleus; dendrites, which spread out like the branches of a tree from the central soma to collect incoming nerve impulses; an axon, which carries nerve impulses away from the soma to a neighbouring neuron; a myelin sheath, which encases the axon, in the way a plastic sheath encases an electrical wire, improving conductivity; and

axon terminals which contact the dendrites of neighbouring neurons.

The axons in each nerve are bundled together like wires in a cable. Long bundles of axons run throughout the body gathering information from and sending instructions to sensory organs, muscles, glands and other structures. Nerve axons may be more than 1 m (3 ft) long in an adult. The longest nerve runs from the base of the spine to the toes.

Neurons can be classified according to the direction in which they carry nerve impulses. Sensory, or afferent, neurons carry information

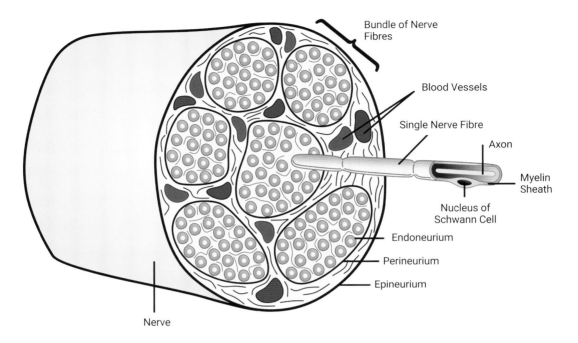

Bundle of Nerve Fibres

Blood Vessels

Single Nerve Fibre

Axon

Myelin Sheath

Nucleus of Schwann Cell

Endoneurium

Perineurium

Epineurium

Nerve

A nerve is composed of many individual nerve fibres, or axons, bundled together in groups called fascicles. Each fascicle is wrapped in a layer of connective tissue called the perineurium and the entire nerve enclosed by the epineurium, a further layer of connective tissue. Within the fascicles the neurons are enclosed within the endoneurium and surrounded by a protective fluid. As well as the nerve fibres, nerves also include the supporting glial cells, such as the Schwann cells that manufacture the myelin sheath around a nerve axon and the blood vessels supplying the nerves with nutrients.

from sensory receptors, such as the eyes or nose, to the central nervous system, changing a physical stimulus, such as light or a scent, into nerve impulses. Motor, or efferent, neurons, transmit instructions from the central nervous system to muscles and glands, activating these structures by means of nerve impulses. Interneurons carry nerve impulses back and forth between sensory and motor neurons within the spinal cord and brain.

Nerve impulses

A neuron that is not actively transmitting a nerve impulse is said to be in a resting state. During the resting state a mechanism called a sodium-potassium pump maintains a difference in charge across the cell membrane of the neuron. Sodium is the principal ion in the fluid outside cells, and potassium is the principal ion in the fluid inside of cells. Differences in concentration create an electrical gradient across the cell membrane, called the resting potential.

An action potential occurs when the neuron receives a chemical signal from another cell that results in a sudden reversal of the electrical gradient across the cell membrane, generating a nerve impulse. A nerve impulse is an all-or-nothing response that is only triggered if the chemical stimulus is strong enough to reach a certain threshold. A neuron responds completely or not at all; a greater stimulus will not produce a stronger impulse.

The transmission of a nerve impulse takes place at a synapse, where an axon terminal meets another cell. The cell sending the nerve impulse is called the presynaptic cell, and the cell receiving it is called the postsynaptic cell. The presynaptic area contains many tiny spherical vessels called synaptic vesicles, packed with chemicals called neurotransmitters. Over a hundred different neurotransmitters are known. When an action potential reaches the axon terminal of the presynaptic cell, it opens channels that allow calcium to enter the terminal. Calcium causes the synaptic vesicles to release their contents into the synaptic cleft, the narrow space between the presynaptic and postsynaptic

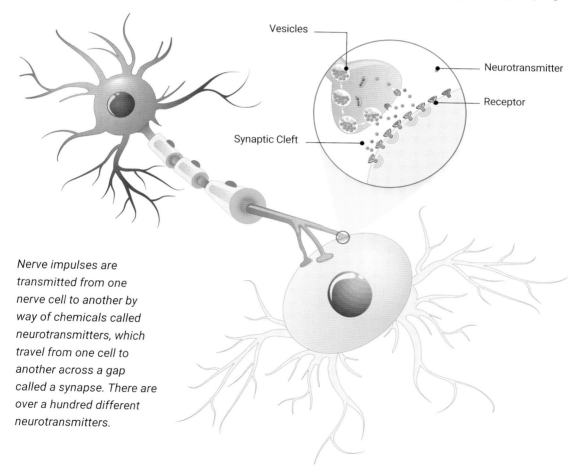

Vesicles

Neurotransmitter

Receptor

Synaptic Cleft

Nerve impulses are transmitted from one nerve cell to another by way of chemicals called neurotransmitters, which travel from one cell to another across a gap called a synapse. There are over a hundred different neurotransmitters.

membranes. The neurotransmitter molecules travel across the synaptic cleft and bind to receptor proteins embedded in the membrane of the post-synaptic cell.

The effect of a neurotransmitter depends mainly on the type of receptors that it activates. A particular neurotransmitter can have different effects on various target cells, exciting one set of cells, for example, while inhibiting others. One of the causes of depression is thought to be an imbalance in the brain of the neurotransmitter serotonin, which normally helps regulate mood. Some antidepressant drugs aim to alleviate depression by regulating serotonin activity.

Nervous tissue

The nervous tissue in the brain and spinal cord consists of grey matter and white matter. Grey

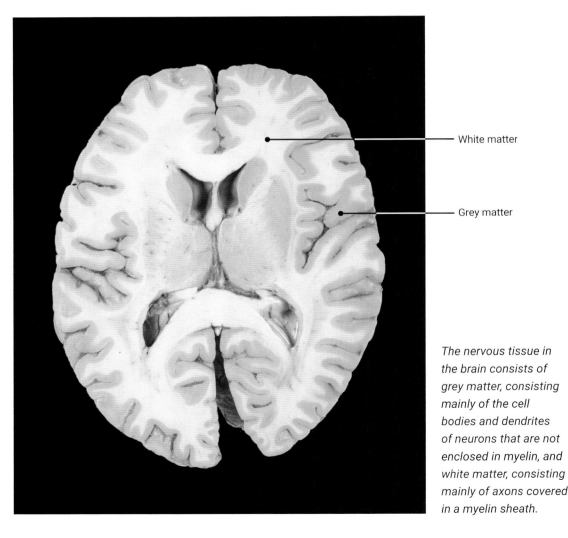

White matter

Grey matter

The nervous tissue in the brain consists of grey matter, consisting mainly of the cell bodies and dendrites of neurons that are not enclosed in myelin, and white matter, consisting mainly of axons covered in a myelin sheath.

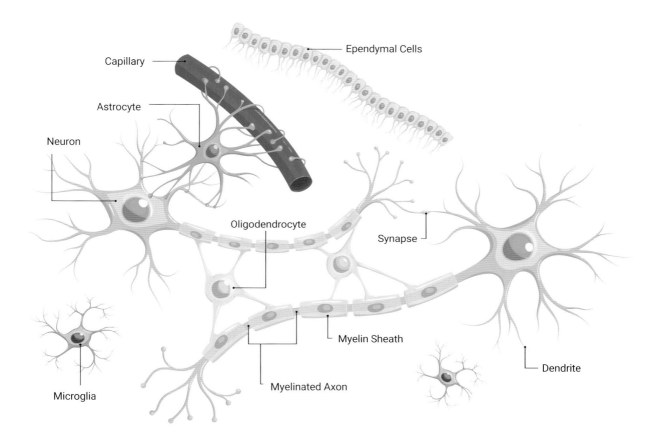

Capillary

Astrocyte

Neuron

Ependymal Cells

Oligodendrocyte

Synapse

Myelin Sheath

Dendrite

Myelinated Axon

Microglia

The different types of glial cell that support the neurons of the central nervous system include ependymal cells which secrete cerebrospinal fluid; astrocytes, the most abundant glial cells linking neurons to their blood supply and forming the blood-brain barrier; microglial cells, a type of macrophage that protect the brain and spinal cord from invading pathogens; and oligodendrocytes which produce the myelin sheaths that insulate the nerve axons allowing signals to propagate more effectively.

matter consists of the cell bodies and dendrites of neurons that lack myelin sheaths. The name is something of a misnomer as it is actually pink, turning grey only after death. White matter consists mainly of axons that do have a myelin sheath, which is what gives them their white colour. The nerves of the peripheral nervous system are also formed of white matter.

In addition to neurons, nervous tissue also includes glial cells, or neuroglia. There are several different types of glial cell, each with a different and vital role to play. Oligodendrocytes

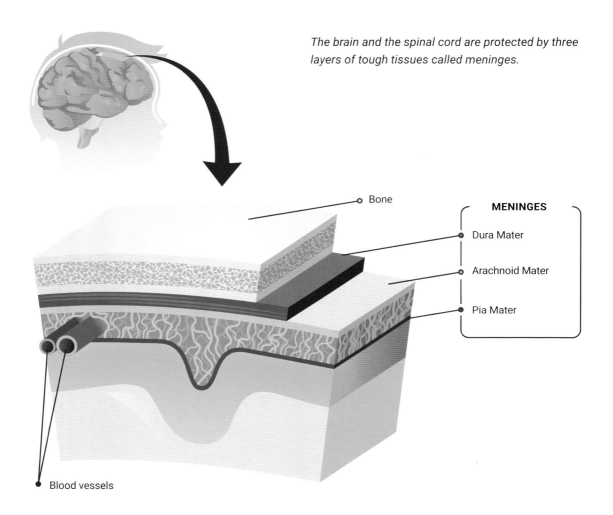

The brain and the spinal cord are protected by three layers of tough tissues called meninges.

Bone

MENINGES

Dura Mater

Arachnoid Mater

Pia Mater

Blood vessels

in the central nervous system and Schwann cells in the peripheral nervous system generate the myelin sheaths, formed from lipids, which increase the speed at which nerve impulses can be transmitted. Other glial cells hold neurons in place, supply them with nutrients, carry out repairs, remove dead neurons, destroy pathogens, and direct axons to their targets. There are roughly equal numbers of neurons and glial cells in the human brain.

CENTRAL NERVOUS SYSTEM

As the central nervous system is one of the most vital parts of the human body, it is no surprise that it is well protected. In addition to being shielded by the bones of the skull and the spinal vertebrae, the brain and spinal cord are protected by three layers of membrane called meninges: the dura, arachnoid and pia mater. Cerebrospinal fluid that flows between the meninges and in fluid-filled ventricles inside the brain cushions and protects

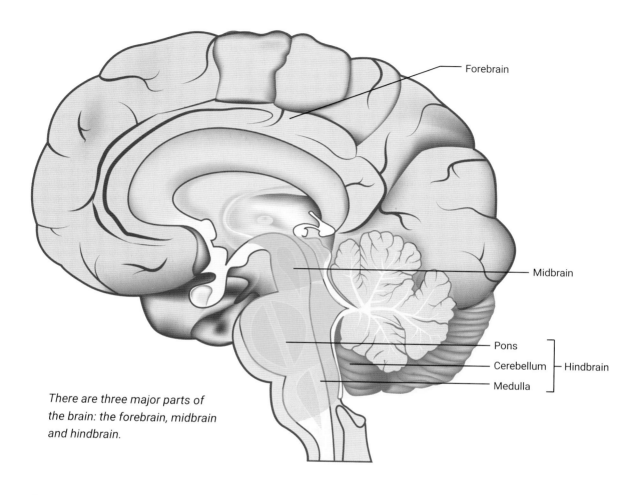

There are three major parts of the brain: the forebrain, midbrain and hindbrain.

Forebrain

Midbrain

Pons
Cerebellum ⎱ Hindbrain
Medulla

the brain and spinal cord. The cerebrospinal fluid also carries nutrients to and removes waste products from the CNS. Cerebrospinal fluid is produced in the ventricles and circulates throughout the CNS before being reabsorbed by the bloodstream.

The brain and spinal cord are further protected. They are isolated from the circulation by the blood-brain barrier, a highly selective membrane that separates the blood in the circulatory system from the extracellular fluid in the CNS, allowing water, glucose, oxygen and some other molecules needed by the brain and spinal cord to cross from the blood into the CNS, but keeping out harmful toxins.

The brain

The human brain contains an estimated 100 billion neurons, each with thousands of synaptic connections to other neurons. It is the control centre of the human body and it works hard at it. It receives and interprets information from the sense organs and sends out instructions for an appropriate response. It co-ordinates voluntary activities such as swimming, walking or washing up as well as unconscious processes such as heart rate and breathing. All of this is energy intensive. About 2 per cent of an adult's body weight is made up by the brain, but it accounts for around 20 per cent of the energy the body consumes, most of which comes from glucose.

The brain has three major regions: the hindbrain, the midbrain and the forebrain, or cerebrum. The hindbrain is the lowest part of the brain. The three areas of the hindbrain are the cerebellum, along with the medulla oblongata and the pons, which together with the midbrain make up the brainstem, connecting the rest of the brain with the spinal cord.

The uppermost part of the brainstem is the

The reticular activating system, a network of nerve pathways connecting different parts of the brain, is responsible for attention, the ability to focus on a task, and for the sleep-wake cycle.

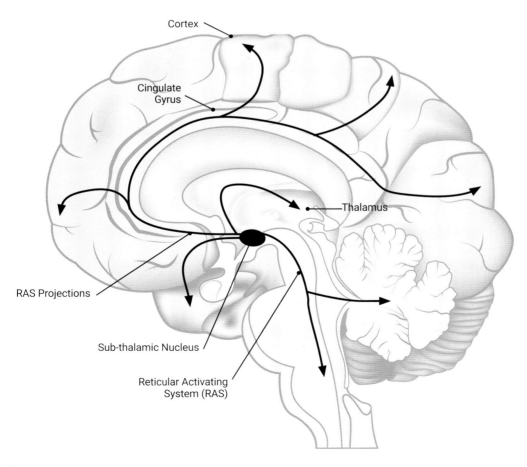

Cortex

Cingulate Gyrus

Thalamus

RAS Projections

Sub-thalamic Nucleus

Reticular Activating System (RAS)

midbrain. It is involved in co-ordinating eye movement. The medulla oblongata is in front of and just below the cerebellum at the very top of the spinal cord. It controls heart rate, respiration rate and blood pressure, along with reflexes such as vomiting and sneezing. The pons sits in front of the cerebellum and above the medulla oblongata. It has several functions, including regulating breathing and sleep cycles and controlling facial expressions.

The reticular activating system (RAS) is a structure within the brainstem that plays a very important role. It is the part of the brain that controls sleep, dreaming and waking. It also acts as a filter for the deluge of information being gathered by the senses, ensuring that only essential data makes its way to the conscious brain. When you first learn to ride a bike, for example, you have to pay close attention to what you are doing, but the RAS gradually automates the process, so you no longer have to think about it.

The cerebellum is located behind the brainstem and just below the cerebrum. It co-ordinates voluntary movements, balance and posture. Data from receptors in the inner ear, muscles and eyes are aggregated by the cerebellum to build up an awareness of the body's position in three-dimensional space. People who have suffered damage to the cerebellum often have difficulty with balance. The cerebellum also plays a major role in learning motor skills like how to ride a bike and plays an important role in memory and learning.

The Cerebrum

The cerebrum is the largest part of the brain. From front to back, the cerebrum is divided into the left and right hemispheres which are connected by a thick bundle of axons, known as the corpus callosum, which transmits signals from one hemisphere to the other. For some reason not yet fully understood, each hemisphere interacts with the opposite side of the body. The left side of the brain receives information from and sends commands to the right side of the body, and the right side of the brain does the same with the left side of the body.

The right and left hemispheres of the cerebrum are similar in shape, and divided into areas called lobes. The frontal lobe (at the front of the head) handles things like attention, behaviour, problem solving, the ability to speak and some types of muscle movements. The parietal lobe (at the top of the head) deals with processing sound and language, numbers and counting, organizing information and decision making. It also deals with touch, temperature and pain signals from sensory organs and is involved in perception, judging distance and the size of objects. The temporal lobe (at the side of the head) is involved in recognition of people and objects and in understanding what others are saying to you. It also forms and retrieves memories and connects memories with emotions. The occipital lobe (at the back of the head) handles the sensory input from the eyes, including the perception of colour and movement. Hidden inside the brain is the insular lobe, located underneath the frontal, parietal and temporal lobes. It handles taste senses and may also process feelings such as compassion and empathy.

Cerebral Cortex

The outer layer of the cerebrum is composed of the billions of neurons and associated glial cells that together make up the cerebral cortex. Just 2 to 4 mm thick, it makes up 40 per cent of the brain's mass. This is where most of the brain's information processing goes on. It is responsible

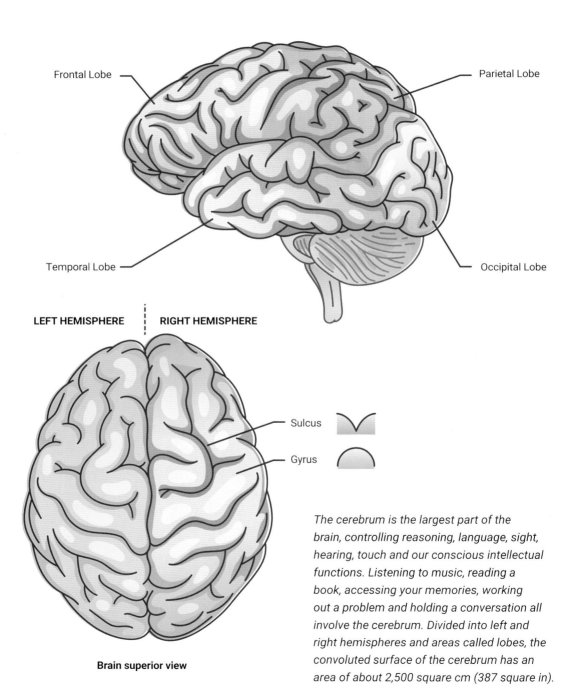

Frontal Lobe

Parietal Lobe

Temporal Lobe

Occipital Lobe

LEFT HEMISPHERE | **RIGHT HEMISPHERE**

Sulcus

Gyrus

Brain superior view

The cerebrum is the largest part of the brain, controlling reasoning, language, sight, hearing, touch and our conscious intellectual functions. Listening to music, reading a book, accessing your memories, working out a problem and holding a conversation all involve the cerebrum. Divided into left and right hemispheres and areas called lobes, the convoluted surface of the cerebrum has an area of about 2,500 square cm (387 square in).

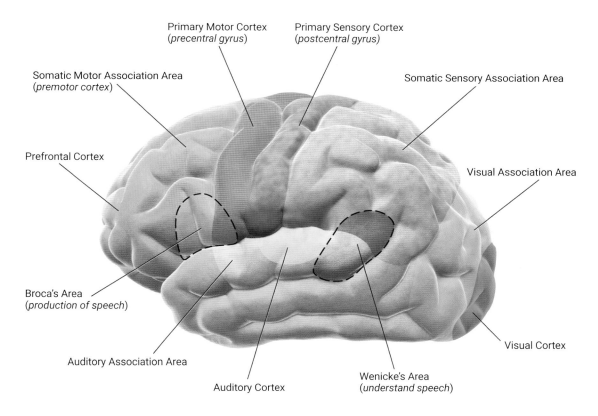

Primary Motor Cortex
(*precentral gyrus*)

Primary Sensory Cortex
(*postcentral gyrus*)

Somatic Motor Association Area
(*premotor cortex*)

Somatic Sensory Association Area

Prefrontal Cortex

Visual Association Area

Broca's Area
(*production of speech*)

Visual Cortex

Auditory Association Area

Wenicke's Area
(*understand speech*)

Auditory Cortex

The cerebral cortex can also be divided into areas that correspond to different functions, such as the visual cortex processing visual information, the primary motor cortex co-ordinating muscle movement and the prefrontal cortex involved in thinking and problem solving.

for controlling much of our conscious thoughts and actions, including language and learning, reasoning, behaviour and personality – as well as directing the movement of the skeletal muscles. If 'you' can be said to be anywhere, it is here.

The limbic system

Several structures located deep within the brain are involved in motivation, emotion, learning and memory as well as being important in communications between the brain and spinal cord. The hypothalamus, thalamus, the hippocampus and amygdala together make up the limbic system. The hypothalamus is located just above the brainstem. It reacts to an array of different signals, both internal and external, including direct input from the brain, light, hormone levels, stress and the presence of pathogens. In response to these inputs the hypothalamus controls certain metabolic

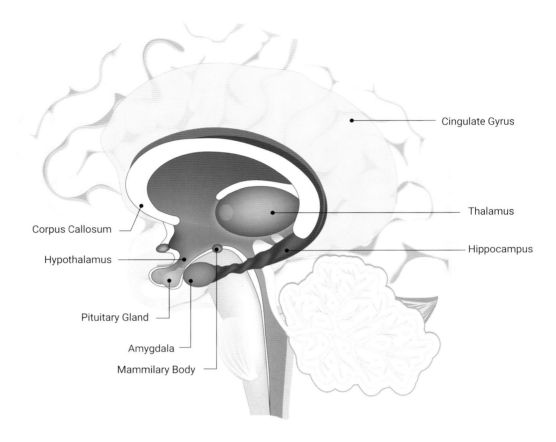

The structures of the limbic system are concerned with learning and memory, motivation and emotion and also regulate essential functions such as body temperature, hunger, thirst and wakefulness.

processes and other activities of the autonomic nervous system, including body temperature, heart rate, hunger, thirst, fatigue, sleep, wakefulness and circadian rhythms. It is also an important emotional centre of the brain.

The hypothalamus mainly controls these functions through its regulation of the pituitary gland, the 'master gland' of the endocrine system. The hypothalamus synthesizes neurohormones called releasing factors that travel to the pituitary gland, stimulating or inhibiting the secretion of pituitary hormones, which control other glands of the endocrine system. It also synthesizes the hormone oxytocin, which stimulates uterine contractions during childbirth, and antidiuretic hormone, which triggers the kidneys to reabsorb more water and excrete more concentrated urine.

The thalamus, located near the hypothalamus, is a major hub for information exchange between the spinal cord and cerebrum, relaying sensory signals to the cerebral cortex and motor signals to the spinal cord. It is also involved in the regulation of consciousness, sleep and alertness.

The hippocampus is embedded deep within the temporal lobe. It plays a major role in learning and memory and in the regulation of motivation and emotion. The amygdala, a small almond-shaped structure, is the part of the brain responsible for the formation and storage of memories associated with emotional events, especially those associated with fear and anxiety. The amygdala plays an essential role in survival as it helps us to learn what is dangerous and should be avoided. It also appears to play a role in aggression and in learning through reward and punishment.

The basal ganglia are a set of structures located near the thalamus and hypothalamus that interact with the limbic system. They receive input from the cerebral cortex and control posture and movement.

The Spinal Cord

Enclosed within the vertebral column of the spine is a long, thin bundle of nervous tissue that extends down from the brainstem, from the medulla oblongata, towards the pelvis. This is the spinal cord. Just 40 to 50 cm (15.5–19.5 in) long and 1 to 1.5 cm (0.4–0.6 in) in diameter, it is one of the most important parts of the human body.

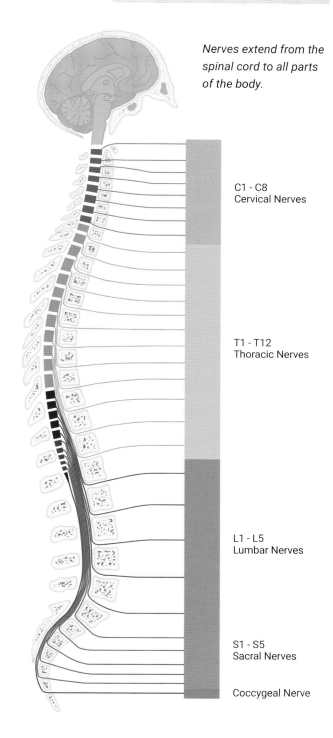

Nerves extend from the spinal cord to all parts of the body.

C1 - C8
Cervical Nerves

T1 - T12
Thoracic Nerves

L1 - L5
Lumbar Nerves

S1 - S5
Sacral Nerves

Coccygeal Nerve

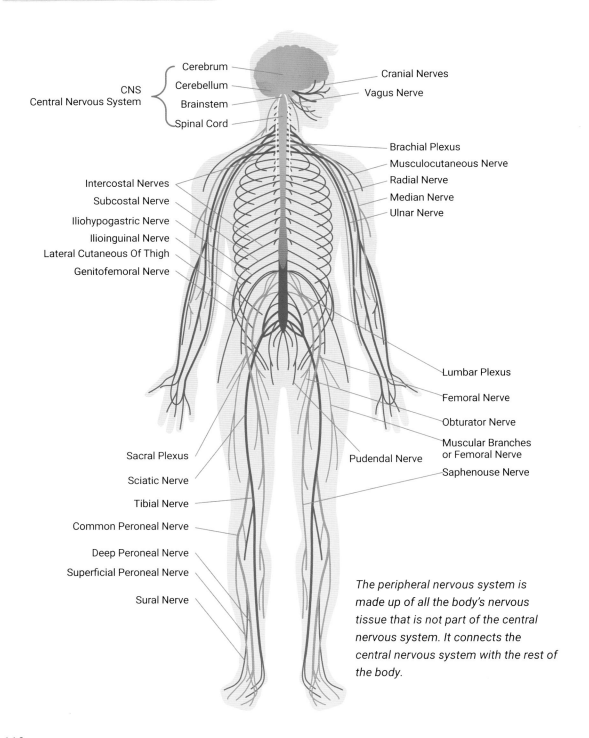

Cerebrum

Cerebellum

CNS
Central Nervous System

Brainstem

Spinal Cord

Cranial Nerves

Vagus Nerve

Brachial Plexus

Musculocutaneous Nerve

Radial Nerve

Median Nerve

Ulnar Nerve

Intercostal Nerves

Subcostal Nerve

Iliohypogastric Nerve

Ilioinguinal Nerve

Lateral Cutaneous Of Thigh

Genitofemoral Nerve

Lumbar Plexus

Femoral Nerve

Obturator Nerve

Muscular Branches
or Femoral Nerve

Saphenouse Nerve

Sacral Plexus

Sciatic Nerve

Tibial Nerve

Common Peroneal Nerve

Deep Peroneal Nerve

Superficial Peroneal Nerve

Sural Nerve

Pudendal Nerve

*The peripheral nervous system is
made up of all the body's nervous
tissue that is not part of the central
nervous system. It connects the
central nervous system with the rest of
the body.*

It has been described as the body's information 'superhighway', carrying messages from the brain to the body and from the body back to the brain. Nerve impulses travel to and from the brain through the spinal cord to specific locations in the body by way of the peripheral nervous system, a complex system of nerves that branch off from the spinal nerve roots.

Thirty-one pairs of nerves and nerve roots extend from the spinal cord. These include:

- eight cervical nerve pairs starting in the neck and running mostly to the face and head;
- twelve thoracic nerve pairs in the upper body that extend to the chest, upper back and abdomen;
- five lumbar nerve pairs in the lower back running to the legs and feet;
- five sacral nerve pairs extend into the pelvis from the lower back.
-

In addition a nerve bundle at the base of the spinal cord called the cauda equina (from the Latin for 'horse's tail') includes nerves that provide sensation to the lower body.

THE PERIPHERAL NERVOUS SYSTEM

All of the body's nervous tissue that is not part of the central nervous system constitutes the peripheral nervous system. Its function is to relay information to and from the rest of the body to the CNS. The PNS lacks the protective blood-brain barrier, bones, meninges and cerebrospinal fluid of the CNS and consequently is more vulnerable to injury and disease. One of the most common causes of peripheral nerve problems is diabetes which can cause numbness or burning and tingling sensations.

The peripheral nervous system (PNS) is divided into the autonomic nervous system and the somatic nervous system.

Autonomic Nervous System

The primary functions of the autonomic nervous system are sensing the internal environment of the body and controlling its involuntary activities. All of the activities that happen inside the body without our conscious awareness of them, such as the processes of the digestive system, are controlled by the autonomic nervous system.

The sensory and motor nerves of the autonomic nervous system run between the central nervous system (especially the hypothalamus in the brain) and internal organs such as the heart, lungs and digestive system organs. Sensory neurons in the autonomic nervous system detect internal body conditions and send messages to the brain, which sends instructions for appropriate responses via motor nerves controlling the contractions of smooth or cardiac muscle, or of glandular tissue. For example, when sensory nerves of the autonomic system detect a rise in body temperature, motor nerves signal smooth muscles in blood vessels near the body surface to undergo vasodilation, and the sweat glands in the skin to secrete more sweat to cool the body.

The autonomic nervous system is divided into the sympathetic division, parasympathetic division and enteric division. The first two both affect the same organs and glands, but in opposite ways. The sympathetic division is responsible for controlling the fight-or-flight response to a perceived threat. For example, the heart rate speeds up, air passages in the lungs become wider to increase uptake of oxygen, blood flow to the skeletal muscles increases, and the activities of

PARASYMPATHETIC
Nervous System

SYMPATHETIC
Nervous System

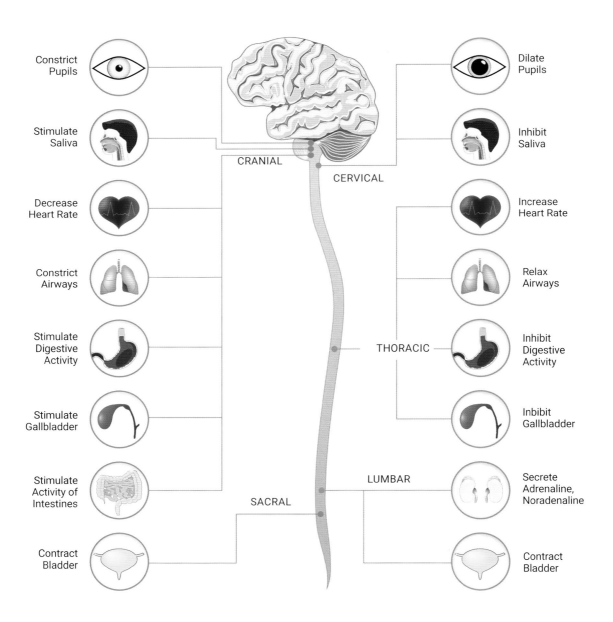

Constrict
Pupils

Stimulate
Saliva

Decrease
Heart Rate

Constrict
Airways

Stimulate
Digestive
Activity

Stimulate
Gallbladder

Stimulate
Activity of
Intestines

Contract
Bladder

CRANIAL

CERVICAL

THORACIC

LUMBAR

SACRAL

Dilate
Pupils

Inhibit
Saliva

Increase
Heart Rate

Relax
Airways

Inhibit
Digestive
Activity

Inbibit
Gallbladder

Secrete
Adrenaline,
Noradenaline

Contract
Bladder

the digestive system are temporarily suspended, all preparing the body for swift action.

Once the threat has passed the parasympathetic division returns the body to normal, slowing down the heart rate, narrowing the air passages in the lungs, reducing blood flow to the muscles, and starting up the digestive system again. The parasympathetic division is also responsible for maintaining the body's homeostatic balance.

The third division, the enteric division, supplies nerves to the organs of the digestive system, controlling the digestive functions.

Somatic Nervous System

The somatic nervous system, also known as the voluntary nervous system, is the division of the peripheral nervous system that is concerned with the voluntary control of body movements via the use of the skeletal muscles. All of the bodily functions we are aware of and can consciously influence are carried out by the somatic nervous system. The somatic nervous system is our link to the external environment. All of the information gathered through the senses is collected by the somatic nervous system for transmission to the brain, where we become consciously aware of our surroundings. If it is necessary to take action in response to the sensory input, such as opening the fridge to find something to eat when you're hungry, or turning up the thermostat when you're cold, the brain's instructions are carried out by the somatic nervous system.

The somatic nervous system consists of both sensory (afferent) and motor (efferent) nerves. It is also responsible for the reflex arc, which involves the use of interneurons to perform reflexive actions. Somatic sensory and motor information is transmitted between the somatic nervous system and the central nervous system through 12 pairs of cranial nerves and 31 pairs of spinal nerves. Cranial nerves in the head and neck connect directly to the brain, transmitting information about smells, tastes, light, sounds and body position. They also control skeletal muscles in the face, tongue, eyeballs, throat, head and shoulders, as well as the salivary glands and swallowing. Four of the 12 pairs of cranial nerves are 'mixed' nerves having both sensory and motor functions.

All of the spinal nerves are mixed nerves. They include sensory nerves in the skin sensing pressure, temperature, vibrations and pain and nerves in the muscles sensing stretching and tension as well as motor nerves stimulating skeletal muscles to contract.

Opposite: The autonomic nervous system monitors the internal environment of the body and controls involuntary responses. It divides into the sympathetic division, which prepares the body for stressful conditions, and the parasympathetic division, which returns things to normal.

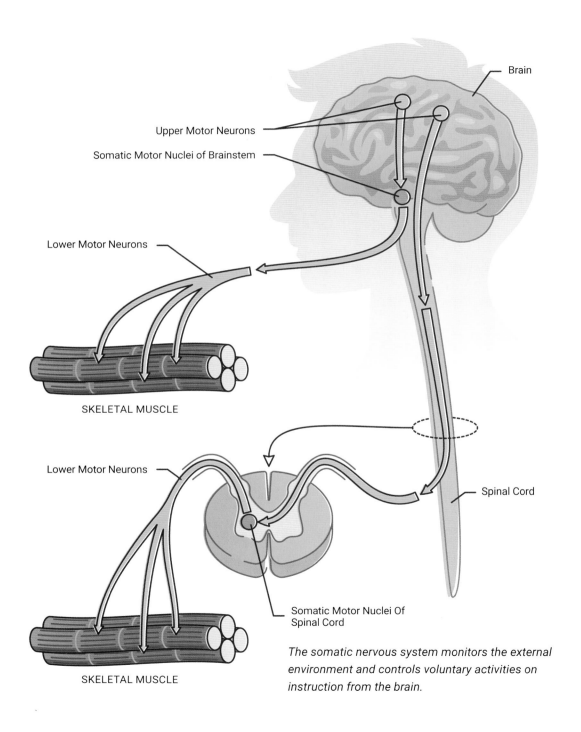

Brain

Upper Motor Neurons

Somatic Motor Nuclei of Brainstem

Lower Motor Neurons

SKELETAL MUSCLE

Lower Motor Neurons

Spinal Cord

Somatic Motor Nuclei Of Spinal Cord

SKELETAL MUSCLE

The somatic nervous system monitors the external environment and controls voluntary activities on instruction from the brain.

REFLEX ARC

A reflex is a quick movement that occurs without conscious thought, such as the immediate withdrawal of your hand if you accidentally touch something sharp. When such a rapid response is called for, the spinal cord acts independently of the brain. Reflexes are so fast because they involve local synaptic connections in the spinal cord, rather than the lengthier process of relaying information to the brain. A sensory receptor responds to a sensation and sends a nerve impulse along a sensory nerve to the spinal cord. The impulse passes to an interneuron and from the interneuron to a motor nerve, which carries the impulse to a muscle which contracts in response, jerking your hand away. This reflex arc, as it is called, requires no input from the brain. You don't have to think about whether or not it's a good idea to take your hand away.

A reflex may only involve one or two synapses. The knee reflex that a doctor tests during a routine check-up is controlled by a single synapse between a sensory neuron and a motor neuron. Synapses with interneurons in the spinal column transmit information to the brain to convey what happened after the event is over. The brain is not involved at all in the reflex action, but it is certainly involved in learning from the experience. It just takes one unfortunate encounter to learn that a rose has sharp thorns!

The spinal cord controls rapid reflex responses without any input from the brain. A sensory receptor sends a nerve impulse along a sensory nerve to the spinal cord where the message passes to an interneuron and from the interneuron to a motor nerve, which carries the impulse to a muscle, which contracts in response.

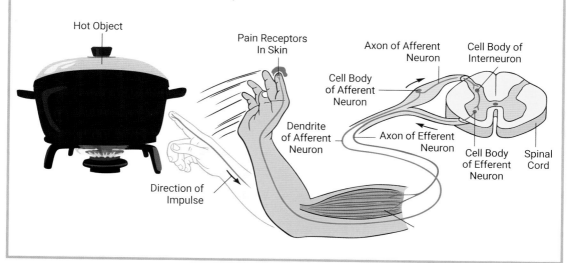

Hot Object

Pain Receptors In Skin

Axon of Afferent Neuron

Cell Body of Interneuron

Cell Body of Afferent Neuron

Dendrite of Afferent Neuron

Axon of Efferent Neuron

Cell Body of Efferent Neuron

Spinal Cord

Direction of Impulse

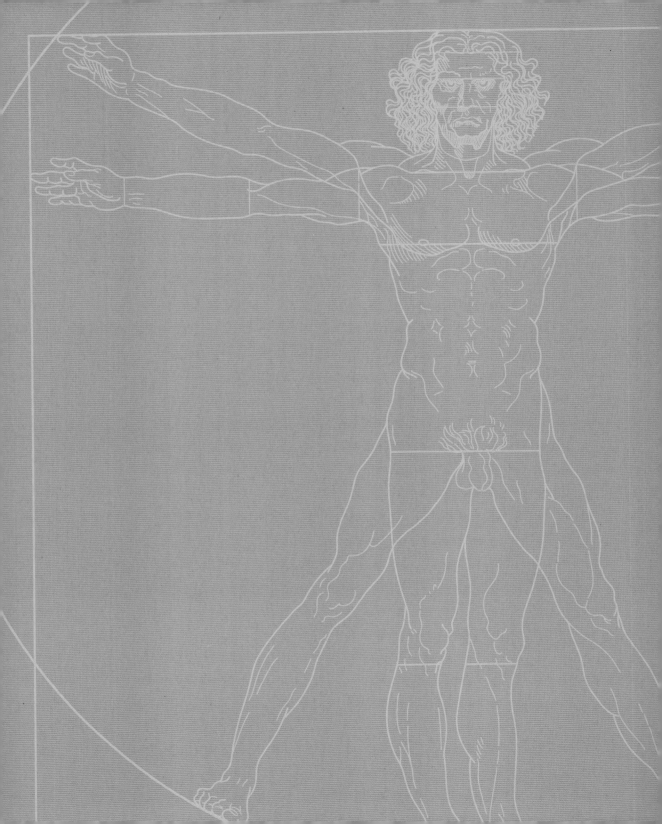

The Senses

The Senses

Senses provide information about the body and its environment. Humans have five special senses: vision, hearing, smell, taste and equilibrium (balance and body position), and general senses, also called somatosensation, responding to stimuli such as temperature, pain, pressure and vibration, together making up the sense of touch. Proprioception (the awareness of the position of bones, joints and muscles), and kinaesthesia (the ability to track limb movement) are part of somatosensation.

A sense can be broadly defined as a group of sensory cells that respond to a specific physical phenomenon. All sensory systems share a common function: to convert a stimulus (such as light, or sound, or the position of the body) into an electrical signal in the nervous system sending information about it to the brain, which receives and interprets that information. This process is called sensory transduction.

The first step in the process of sensation is the activation of sensory receptors by a stimulus, such as light falling on the receptor cells of the retina. Sensory cells are specialized according to the type of stimulus they detect. Receptors in the eye are unaffected by sound, and receptors in the ear do not react to light for example. These signals are detected by different types of sensory receptor in the form of specialized protein molecules.

Detecting light involves proteins called opsins.

Below and Opposite: Many of the body's sensory receptors rely on specialized protein molecules to detect various stimuli. Light detection involves proteins called opsins which change shape; motion-detecting proteins allow charged particles to pass through when activated; chemical-sensitive proteins are activated by interacting with particular molecules; temperature sensors change shape with changing temperatures.

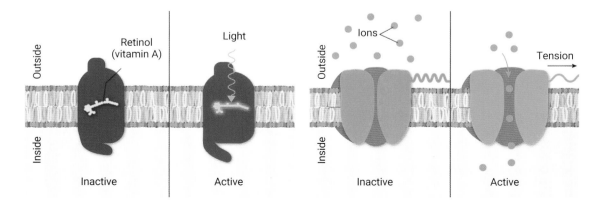

Outside — Inside

Retinol (vitamin A)   Light

Inactive   Active

Ions   Tension

Inactive   Active

All opsin proteins have a molecule of vitamin A embedded within them. When a photon (the smallest detectable unit of light) strikes an opsin molecule its energy is absorbed by the protein, temporarily changing its shape. This is the first step in a signalling chain that ultimately reaches the brain.

Most motion-sensitive proteins extend through the membrane of specialized motion-detecting cells. They respond to mechanical signals in the form of movement, stretching or vibration. Activation causes the channels to open, allowing ions (charged molecules) to pass through.

Chemical-sensitive protein receptors are activated through a lock-and-key mechanism, where a molecule with a specific shape attaches to the receptor. Temperature sensors are modified versions of chemical sensors. Instead of responding to interactions with other molecules, they change shape in response to temperature. A few temperature-sensitive proteins are activated by chemicals as well as temperature. For example, this is why biting into a chilli produces a perception of heat in the mouth.

The distance over which a sensory receptor can respond to a stimulus is called its receptive field. Taste, for example, requires direct contact between the stimulus and the receptor, but vision on the other hand operates over distances as great as that separating us from a faraway star.

How we interpret the signals received from the senses is called perception. This takes place in the brain. All sensory signals, except those from the olfactory system, are transmitted through the central nervous system and are routed to the thalamus which acts as a relay station for sensory signals, passing them on to the appropriate area of the cortex. The brain will often combine information from a number of senses, a process called sensory integration which takes place in the brain's association areas.

EYES AND SIGHT

The ability to sense light and form images of the world around us is one of the most important of the senses. The sensory organ responsible for light gathering is the eye, which gathers and focuses light to form an image. It then converts that image into nerve impulses that travel to the brain for processing, a task to which more than 50 per cent of the cerebral cortex is devoted.

Light enters the eye through the cornea, a clear outer layer that both protects the eye and

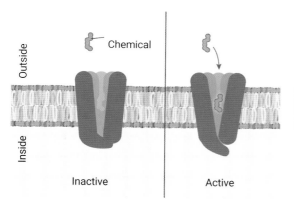

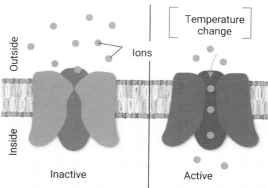

helps to focus the light by refracting it. Around 80 per cent of the eye's focusing power comes via the cornea. Inside the cornea is the hard, white outside of the eyeball called the sclera. The muscles that move the eye are attached to the sclera. The light travels into the interior of the eye through an opening called the pupil, the size of which is controlled by the surrounding iris, which adjusts the opening of the pupil according to how bright the light is – narrowing the pupil in bright light and widening it in dim light. Pigmentation of the iris is what determines eye colour. A semi-gelatinous fluid called aqueous humour fills the space between the cornea and the iris and helps to maintain the shape of the eye.

On the other side of the pupil the light passes through the lens, which refracts the light, focusing it as an inverted image on the retina at the back of the eye. Around 20 per cent of the eye's focusing power is provided by the lens. The curvature of the lens is controlled by the ciliary muscles which accomplish this by contracting to lessen tension on the lens so it becomes rounder, allowing it to bend light rays more, or relaxing for the opposite effect. Ciliary muscle is smooth muscle and so not under voluntary control. Although it only supplies a small proportion of the focusing power the lens is important because it allows us to accommodate – the ability to focus on objects both near and far. Between the lens and the retina,

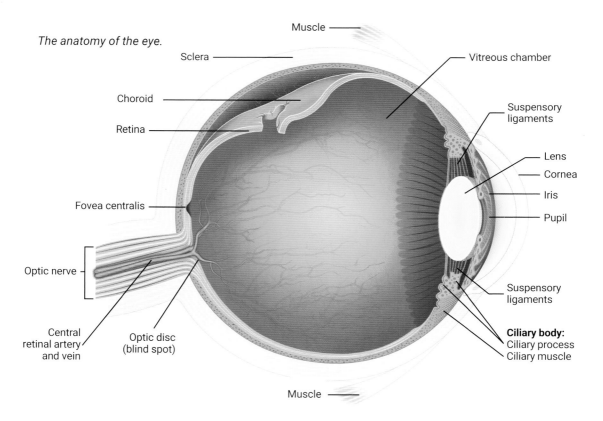

The anatomy of the eye.

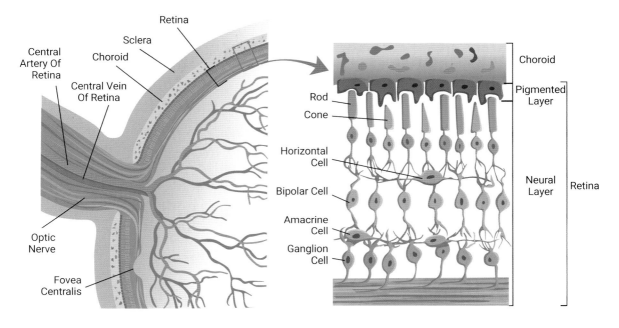

Labels (left illustration): Central Artery Of Retina, Central Vein Of Retina, Choroid, Sclera, Retina, Optic Nerve, Fovea Centralis

Labels (right illustration): Rod, Cone, Horizontal Cell, Bipolar Cell, Amacrine Cell, Ganglion Cell, Choroid, Pigmented Layer, Neural Layer, Retina

The structure of the retina. As shown in the right-hand illustration, light enters from the bottom and has to pass through layers of supporting cells before reaching the light-sensitive rods and cones. The choroid layer behind the retina carries the blood vessels that supply the retina with nutrients and remove wastes.

the gelatinous vitreous humour also functions to maintain the eye's shape.

The retina is a sheet of neurons, an extension of the central nervous system, that contains two types of photoreceptors: rod and cone cells, so named because of their shape. Embedded in the rods and cones are molecules, called photopigments, that absorb certain wavelengths of light. Rods greatly outnumber cones (there are around 120 million rods to six million cones in the average retina) and are found in all areas of the retina with the exception of the very centre. Rod cells are particularly sensitive to low levels of light and in fact do not work in bright light. After being saturated with light in bright conditions

the rods take seven to ten minutes to adapt to dim light, which is why we slowly begin to make out objects in a darkened room as our 'night vision' comes into play. We cannot make out fine detail using rods.

The centre of the retina, an area called the macula, is where the cone cells are most concentrated. At the centre of the macula is the fovea, an area of densely packed cones with no rods present. The cones are sensitive to light of different colours, giving us colour vision and the ability to perceive fine detail. We have three types of cone cell – blue, green and red – each sensitive to a specific wavelength of light. The combined stimulus from the different cone cells

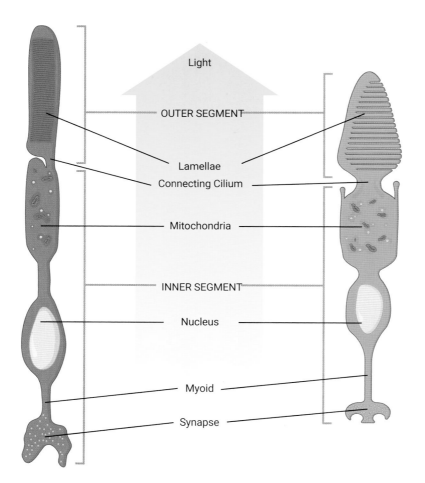

Light

OUTER SEGMENT

Lamellae

Connecting Cilium

Mitochondria

INNER SEGMENT

Nucleus

Myoid

Synapse

Both rods and cones have an outer segment, an inner segment and a synaptic region. Although similar in overall structure, rods are more elongated and narrower than cones. The outer segment is where the photoreceptor molecules (rhodopsin for rods; iodopsin for cones) are found. The inner segment is where the cell's organelles, including the mitochondria and nucleus, are located. The synaptic region is where the rod or cone relays information to the neurons in the retina for transmission along the optic nerve to the brain.

is perceived as a specific colour. A person with normal colour vision can differentiate between hundreds of thousands of different colours, hues and shades.

Both rods and cones convert the light that strikes them to nerve impulses which travel to the optic nerve which has about one million axons. In the area on the retina occupied by the axons of the optic nerve there are no photoreceptors, creating a 'blind spot'. Normally we don't perceive this blind spot because the brain fills the gap by copying whatever happens to be around it. Behind the retina is a black layer called the

pigment epithelium where the photopigments are replenished.

A group of axons from the optic nerve of each eye crosses over to join the opposite optic nerve at the optic chiasm, which means that each side of the brain receives visual information from both eyes. After the chiasm, the optic nerve axons go to one of three areas, two of which are located in the midbrain and one is in the thalamus. The information going to the midbrain produces eye movements and reflex reactions in the pupil. These are controlled by the autonomic nervous system and not

Crossover of the optic nerves from each eye at the optic chiasm ensures that both left and right hemispheres of the brain receive information from both left and right eyes. This information is organized by the hypothalamus before onward transmission to the visual cortex for processing.

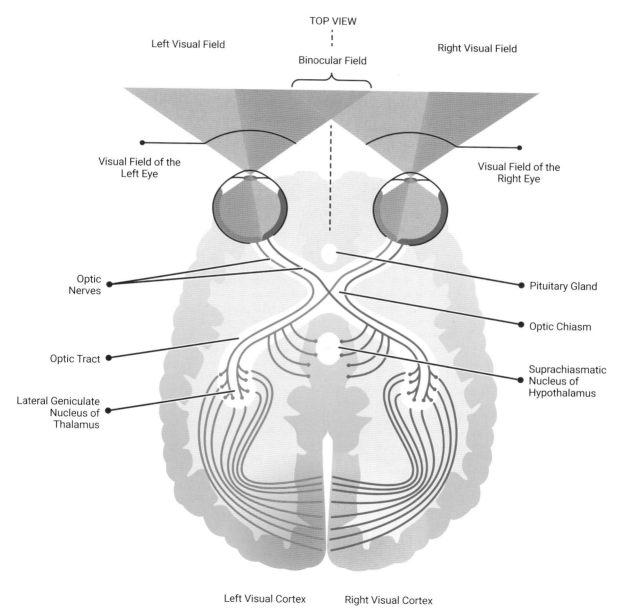

TOP VIEW

Left Visual Field

Binocular Field

Right Visual Field

Visual Field of the Left Eye

Visual Field of the Right Eye

Optic Nerves

Pituitary Gland

Optic Chiasm

Optic Tract

Suprachiasmatic Nucleus of Hypothalamus

Lateral Geniculate Nucleus of Thalamus

Left Visual Cortex   Right Visual Cortex

perceived consciously. In the thalamus, axons transmit signals to neurons in a structure called the lateral geniculate nucleus where information is processed and then carried to the primary visual cortex in the occipital lobe of the cerebrum. The visual cortex, responsible for interpreting the information from the eyes, is the largest system in the human brain. This is where the real business of 'seeing' takes place.

EARS AND HEARING

All sounds are produced by vibrations, in other words by something moving back and forth in a regular or random manner. In order for us to hear these sounds the vibrations must be converted into neural impulses for transmission to the brain. This conversion is accomplished by the ear.

The human ear has three distinct sections: the outer ear, the middle ear and the inner ear. The outer ear consists of the pinna, the only part of the ear that is visible, the external auditory canal, or meatus, and the tympanic membrane, or eardrum. The pinnae, one on each side of the head, are basically skin-covered flaps of cartilage. They collect and focus sound, allowing us to judge its direction and enhancing its intensity.

The central part of the pinna, called the concha, directs sound along the 2.5 cm (1 in) length of the external auditory canal to the tympanic membrane, causing it to vibrate. Behind the eardrum is an air-filled cavity, in which are located the smallest bones in the human body.

The three interlocking bones of the middle ear are called ossicles. Their function is to transmit

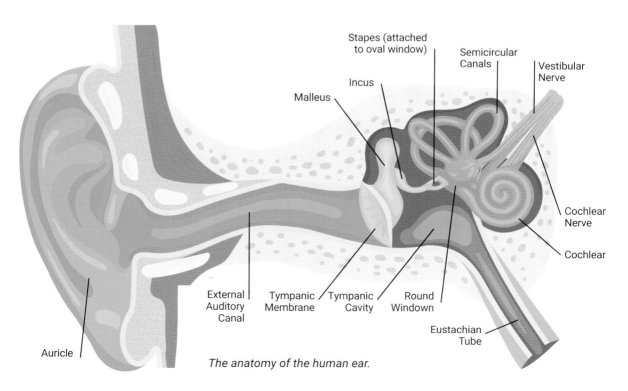

The anatomy of the human ear.

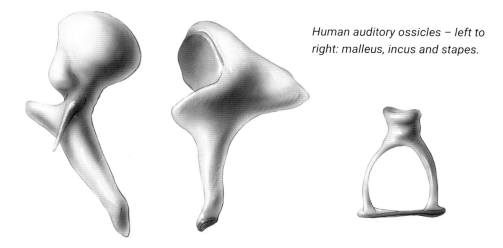

Human auditory ossicles – left to right: malleus, incus and stapes.

the vibrations of the eardrum onwards to the inner ear. The malleus (or hammer) is attached to the eardrum and to the incus (anvil) which has a flexible connection to the stapes (stirrup). The flat-bottomed part of the stapes, called the footplate, is connected in turn to a second membrane, called the oval window. The vibrating eardrum moves the malleus and incus which tug the stapes back and forth against the oval window, on the other side of which is the inner ear. The smallest muscle in the body, called the stapedius muscle, is attached to the stapes. It acts to dampen large vibrations caused by loud noises, stabilizing the stapes and protecting the oval window.

The movements of the oval window set up waves in fluid-filled ducts within one of the most remarkable organs in the human body – the cochlea, so named because of its striking resemblance to the shell of a snail. A hollow tube with bony walls wraps around a central bony structure called the modiolus. Unwound, the tube is around 32 mm (1.25 in) in length and 2 mm in diameter. It is divided into three chambers: the scala vestibuli, scala tympani and the central scala media, or cochlear duct. Scala, Latin for stairway, describes how the scalae wrap around the cochlea like a spiral staircase.

The basilar membrane lies between the scala tympani and the scala media. The base of the basilar membrane is about five times narrower and around a hundred times stiffer than its apex. Sounds of different frequency cause the membrane to vibrate more strongly at different points. High-frequency sounds produce the greatest vibrations near the base of the membrane, and low frequencies produce the greatest vibration near its apex.

On top of the basilar membrane, within the cochlear duct, lies the organ of Corti. This houses sensory receptor cells, known as hair cells, which project into the jelly-like tectorial membrane which lies above it. It is the job of these hair cells to convert the mechanical energy of the basilar membrane into electrical signals for transmission to the brain.

There are about 16,000 hair cells in the human ear running the length of the basilar membrane. When sound stimulates a point on the membrane the hair cells at that site are stimulated by the force that this movement creates. Different groups

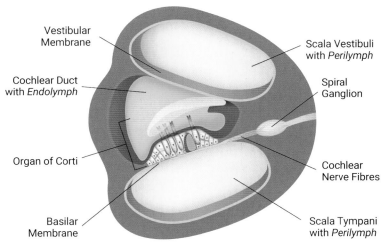

A cross-section of the cochlea.

Vestibular Membrane

Scala Vestibuli with *Perilymph*

Cochlear Duct with *Endolymph*

Spiral Ganglion

Organ of Corti

Cochlear Nerve Fibres

Basilar Membrane

Scala Tympani with *Perilymph*

of hair cells therefore only respond to different frequencies of sound.

The top of each hair cell is topped with a number of filaments called stereocilia. As the basilar membrane vibrates back and forth, the stereocilia are bent at an angle against the tectorial membrane. The bending of the stereocilia produces an electrical response in the hair cells, which in turn trigger the neurons in the auditory, or cochlear, nerve which carries sensory information to the brain. The hair cells have to respond extremely rapidly. The highest frequency detectable by the human ear is around 20,000 Hz, which effectively means that the hair cells must be

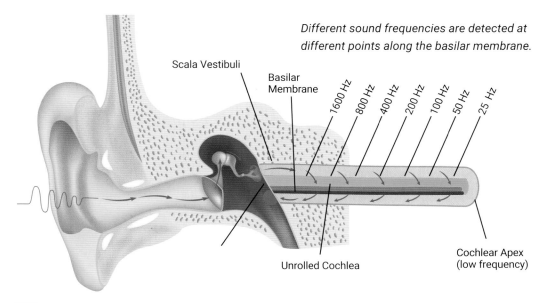

Different sound frequencies are detected at different points along the basilar membrane.

Scala Vestibuli

Basilar Membrane

1600 Hz

800 Hz

400 Hz

200 Hz

100 Hz

50 Hz

25 Hz

Unrolled Cochlea

Cochlear Apex (low frequency)

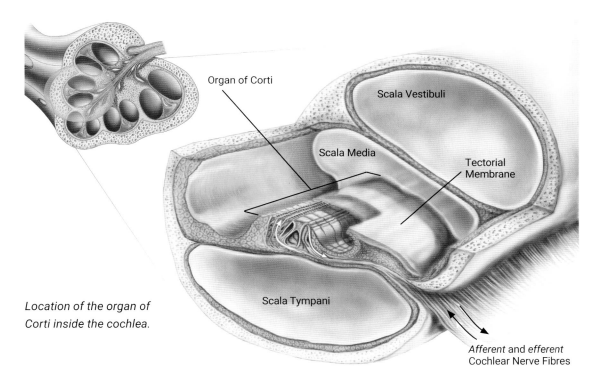

Organ of Corti

Scala Vestibuli

Scala Media

Tectorial Membrane

Scala Tympani

Location of the organ of Corti inside the cochlea.

Afferent and *efferent* Cochlear Nerve Fibres

able to turn current on and off 20,000 times every second. The sensitive hair cells are susceptible to permanent damage if exposed to prolonged loud noises, resulting in hearing loss.

A SENSE OF BALANCE

As well as allowing us to hear, the ear is also essential to our sense of balance: the organ of balance, called the vestibular system, or apparatus, is found inside the inner ear. Housed within the bony labyrinth, a system of tunnels in the hardest part of the temporal bone, the two structures that make up the vestibular apparatus are the vestibule and the semi-circular canals.

Inside the vestibule, which is about 3 to 5mm across, are two membranous sacs, called the saccule and utricle, together known as the otolithic organs. Inside the otolithic organs are patches of

hair cells topped by calcium carbonate crystals called otoconia. Gravity bends the hair cells, due to the weight of the otoconia and this bending produces nerve signals which travel along the vestibular nerve to the brainstem. The otolithic organs detect acceleration, when an elevator descends, for instance, or an aircraft gathers speed for take-off. The utricle is most sensitive to tilt when the head is upright, while the saccule is most sensitive to tilt when the head is horizontal.

The three semi-circular canals, called the superior, posterior and lateral canals, range from 12 to 22mm in length and lie at right angles to one another with the ends of each canal opening out into the vestibule.

At the swollen base of each fluid-filled canal is a crista. This holds hair cells, which press into a jelly-like cupola. These hair cells act as sensory

Electron microscope images of hair cells in the inner ear.

receptors. Movement causes displacement of the fluid in the canals corresponding to the direction of movement, causing the hairs to bend. Each of the three semicircular canals is responsible for a specific direction of head movement: one of the canals responds to the head tilting upwards or downwards, one to it tilting to the right or to the left, and the third to it turning sideways. Whenever we turn our head, the inner ear turns along with it. This results in an electrical signal in the nerve cells connected to the hair cells, which is then carried via the vestibular nerve to the base of the brain for processing and then sent on to other organs that need this information, such as the eyes, joints or muscles. This information allows us to keep our balance and contributes to the sense of proprioception, knowing what position our body is in (see page 136).

Found inside the otolithic organs are patches of hair cells topped by calcium carbonate crystals called otoconia. Gravity bends hair cells due to the weight of the otoconia and this bending produces nerve signals which travel along the vestibular nerve to the brainstem.

THE NOSE AND SMELL

The sense of smell (along with that of taste) is a chemical sense. Both olfactory (odour) and taste receptors are chemoreceptors, sending nerve impulses to the brain where they are interpreted as smells and tastes. Chemoreceptors in the nose allow us to detect volatile chemicals, or odorants, suspended in the air. Often these odours can be detected long before their source becomes apparent. Current research indicates that the average human can distinguish as many as a trillion distinct odours, ranging from the sweet aroma of fresh fruit to the less pleasant smell of rotting meat. All of these odours can be classified into six major groups: floral, fruit, spicy, resin, burnt and putrid.

The sense of smell begins when these volatile chemicals in the air enter through the nostrils and into the nasal cavity, an air-filled space just behind the front of the skull. The odorants dissolve in the olfactory epithelium, a mucus-covered area towards the top rear of the nasal cavity. Embedded within the olfactory epithelium are the olfactory receptor neurons, the sensory neurons for the olfactory system. The average human has somewhere between 6 and 20 million olfactory receptor neurons. By contrast, dogs may have a billion or more. Also found within the olfactory epithelium and surrounding the olfactory receptor neurons are supporting cells, which help to dispose of dead cells and deal with pollutants, and basal cells which replace the olfactory receptor neurons as necessary.

Olfactory receptor neurons have a single dendrite with many hair-like cilia that stretch into the mucus on the surface of the olfactory

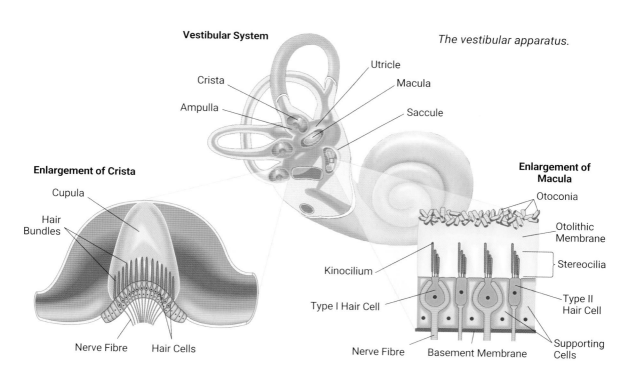

Vestibular System

The vestibular apparatus.

Crista
Ampulla

Utricle
Macula
Saccule

Enlargement of Crista

Cupula
Hair Bundles
Nerve Fibre
Hair Cells

Enlargement of Macula

Otoconia
Otolithic Membrane
Stereocilia
Kinocilium
Type I Hair Cell
Type II Hair Cell
Nerve Fibre
Basement Membrane
Supporting Cells

epithelium, increasing the area available to react with any odorants. They are the only nerve cells in the body that are directly exposed to the outside world in this way. When odorants dissolve in the mucus they interact with odorant receptor proteins on the cilia of the olfactory receptor neuron, a process that leads to the generation of a nerve impulse. Which axons are activated depends upon which receptor protein is targeted. Each cell produces only one type of receptor protein. Humans have around 400 different types of receptor proteins; other animals, for example mice, can have upwards of a thousand.

Although there are only 400 different receptors humans can, as we have said, distinguish a far greater number of different odours. Researchers believe this is possible because some odorant molecules can activate more than one receptor. Think of the receptors as being analogous to the keys on a piano. Depending on which ones are 'played' the result might be a pleasant odour (a melodious chord), or an unpleasant stench (a crashing discordance).

The axons of these neurons, which together make up the olfactory nerve, travel through a sheet of bone called the cribriform plate before reaching the olfactory bulb. Here, the initial processing of odour information takes place in structures called glomeruli. Each glomerulus, which is specific to one type of receptor, houses tens of thousands of axons from the receptor cells as well as dendrites from other neurons connecting the olfactory bulb to the other regions of the brain. Signals from the glomeruli travel directly to the olfactory cortex and from there to the frontal cortex and the thalamus. This is a different path from most other sensory information, which goes first to the thalamus and then to the cortex. Olfactory signals also travel directly to the amygdala, an area of the brain that deals with memory and emotions.

THE TONGUE AND TASTE

Taste, like smell, depends on the ability to detect certain chemicals. It is an important sense in that, along with the sense of smell, it allows us to evaluate whether or not something is good to eat. Taste receptors are bundled in clusters on the surface of the tongue called taste buds. Taste buds themselves are too small to see without the aid of a microscope, but most are concentrated in small peg-like structures on the tongue called papillae which can easily be seen on the tongue's surface.

There are three different types of papillae based on their shape, each type located in a specific area of the tongue. Each taste bud is composed of between 50 and 150 taste receptor cells bundled together like a cluster of bananas, arranged so that their tips form a small taste pore, through which hair-like microvilli, which hold the taste receptors, extend from the taste cells.

Not all of the tongue's taste receptors sense the same kinds of chemicals, or tastes. Humans have five kinds of taste buds – salty, sweet, sour, bitter and umami. It was originally believed that each type of taste bud was found only in specific areas of the tongue, but it is now known that each taste can be detected anywhere on the tongue's surface, although some may be concentrated more in one area than another.

Three cranial nerves link the tongue to the brainstem at the medulla. From there, information is sent to the thalamus and on to the gustatory cortex where information processing takes place.

Taste receptor cells are mostly located on the tongue, but there are other regions of the mouth and throat, including the palate, pharynx and epiglottis, that are sensitive to food and also play

The sense of smell is enabled by the olfactory nerve, the shortest sensory nerve in the body, leading from the brain to the olfactory bulb. Specialized neurons in the bulb travel to the upper inside of the nasal cavity where they react to odours in the air.

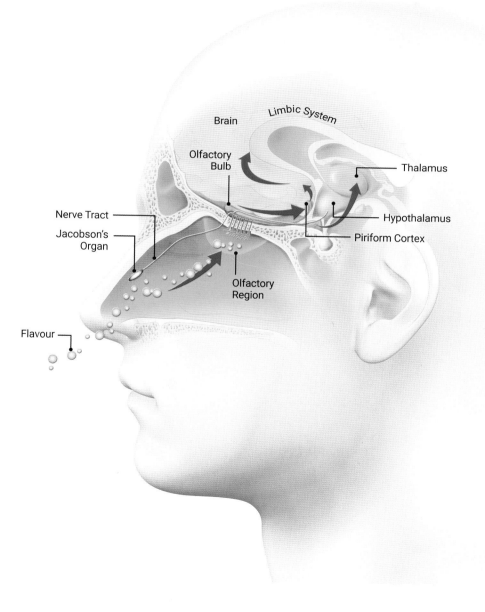

Limbic System

Brain

Olfactory Bulb

Thalamus

Nerve Tract

Jacobson's Organ

Hypothalamus

Piriform Cortex

Olfactory Region

Flavour

a role in taste perception. The olfactory system is intimately linked to the sense of taste as well. What we understand as the sensation of flavour is a combination of taste and smell. As we chew our food, food odorants travel from the mouth to the nasal passage where they are detected by the sensory cells in the nose. The combined signals of taste perception from the tongue and odour perception from the nose are together interpreted by the brain as flavour or taste.

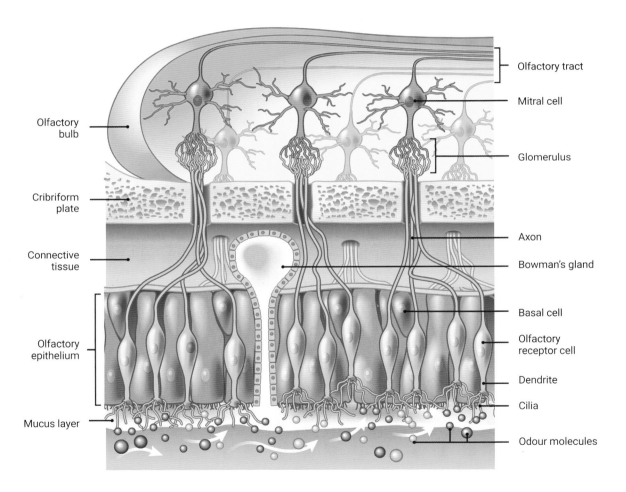

Olfactory tract

Mitral cell

Olfactory bulb

Glomerulus

Cribriform plate

Connective tissue

Axon

Bowman's gland

Basal cell

Olfactory receptor cell

Olfactory epithelium

Dendrite

Cilia

Mucus layer

Odour molecules

Odour molecules dissolve in the mucous layer produced by the Bowman's glands and bind with the olfactory receptor cells. Axons from the receptor cells project through the cribriform plate, linking with dendrites of mitral cells in spherical structures called glomeruli. Axons from the mitral cells form the olfactory nerve travelling to the brain.

TASTE TYPES

Sweet is pleasant and indicates the presence of carbohydrates in food which are a good source of energy.

Salty detects the presence of salt (sodium chloride), a crucial part of the diet.

Sour also indicates the presence of salt, but also of acids. It can be pleasant in small amounts but becomes unpleasant if there is too much. It can be an indicator of food that is spoiled or harmfully acidic.

Bitter is generally an unpleasant taste. Plants sometimes produce bitter toxins as a defence against being eaten.

Umami, or savoury, was first described more than a century ago by a Japanese researcher and is now generally recognized as the fifth taste. Cheese, meat and soy sauce produce this generally pleasant taste. It signals the presence of amino acids and proteins in food, essential components of the diet.

Taste receptor cells are located in bumps on the surface of the tongue called papillae which can take a variety of different forms.

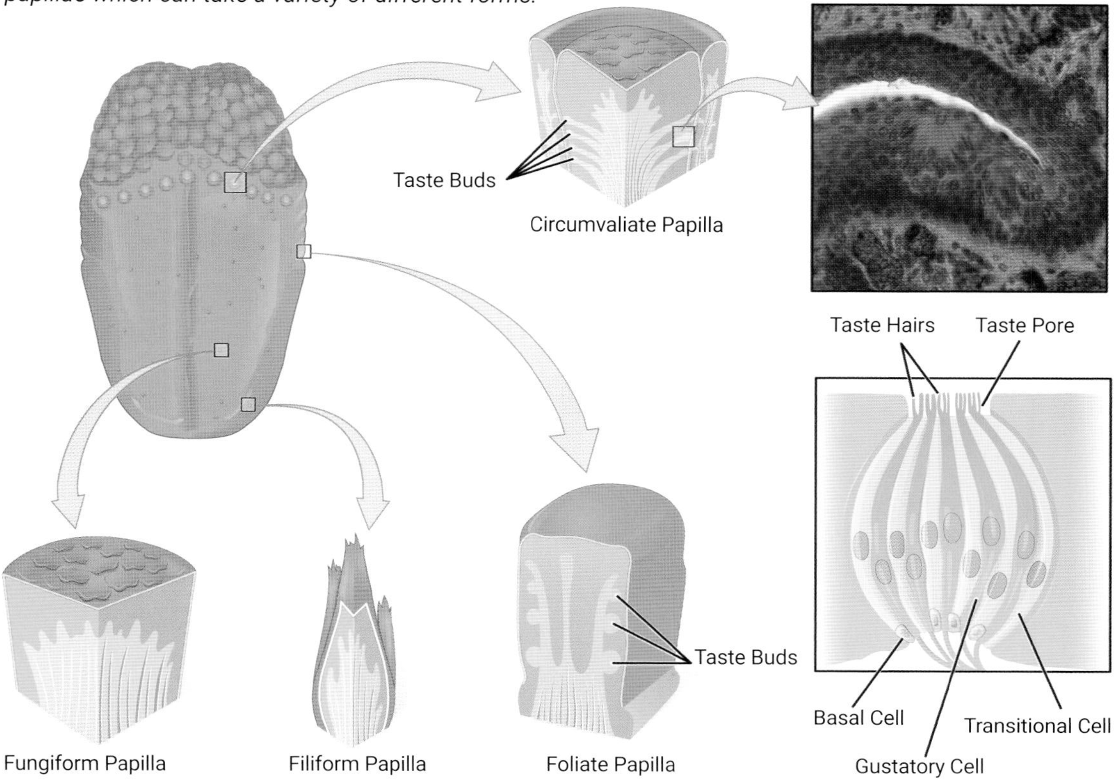

Taste Buds

Circumvaliate Papilla

Taste Hairs   Taste Pore

Basal Cell   Transitional Cell

Gustatory Cell

Taste Buds

Fungiform Papilla   Filiform Papilla   Foliate Papilla

THE SENSE OF TOUCH

One of the main ways in which we experience the world around us is through our sense of touch. The primary sensory organ for touch is the skin, the largest organ in the body. What we commonly think of as 'touch' involves more than one kind of stimulus. The skin's sensory system is a complex one with a variety of sensors that give rise to a range of different sensations that include pain, itching, heat, vibrations, light pressure or a gentle caress.

Each sensor has physical properties that allow it to detect specific signals. Some are known as mechanical sensors, embedded within structures that transmit or amplify movement such as pressure or vibration. Others have sensors embedded in the membranes of cells that change shape in response to a temperature change or to the presence of a chemical. Whether it is temperature, movement or a chemical that is being detected, this information has to be converted into an electrical impulse to be transmitted along neurons to the brain. In order for this to happen channels on the sensory cells open and close, letting ions pass in and out which start off the impulse.

Merkel cells, which are concentrated in our fingers and lips, are the receptors that detect a gentle touch. They are particularly important in the development of fine motor skills as they allow us to detect light pressure and the edges and textures of objects. Without Merkel cells, simple tasks such as using a knife and fork or buttoning a shirt would be difficult. The number of Merkel cells in our skin decreases with age, which may explain why older people are less sensitive to touch.

Merkel cells sit just under the surface in hairless skin; in hairy skin they surround the base of fine hairs. Merkel cells are not neurons. They communicate with the nervous system by producing a neurotransmitter called norepinephrine which activates nearby neurons, triggering them to send signals to the brain.

Ruffini corpuscles are found deep under the skin and inside joints. They are long, thin structures that squeeze the nerve endings inside them when they are stretched. They play a part in proprioception (knowing where our body parts are in space) and help us to feel and manipulate objects in the hand, such as a screwdriver or a fork.

Pacinian corpuscles also sit deep beneath the skin. They respond to rapid changes in pressure. A sudden poke will cause the large, squashy corpuscle to deform, triggering the nerve ending inside it. They are important in helping us feel the texture of an object.

Meissner's corpuscles sit just below the skin surface of hairless skin, such as that on the soles of the feet, palms of the hands and the lips. They consist of a stack of fat, slippery layers, through the centre of which a nerve ending runs, branching out between the layers. Movement across the skin causes the layers of the corpuscle to slide across each other, triggering the nerve ending.

Lanceolate nerve endings surround the bases of the hairs that cover most of our skin surface. If something brushes against the hair it triggers these nerve endings, transmitting information about whatever has moved across the skin.

Humans have at least six types of temperature-sensitive channels, each sensitive to a different temperature range. These TRP (pronounced 'trip') channels each have different nerve endings responding to cold, cool, warm or hot temperatures. Some of the proteins involved in TRP channels will also respond to chemical signals, for instance chemicals in mint will activate cold-sensing TRP channels, which is why we talk about something minty having a 'cool' taste.

Some nerve endings in the skin are sensitive to chemical signals from compounds such as histamine, a chemical that is released by the immune cells of the body in response to an insect bite, or a sting from a nettle. When these nerve endings are triggered, we perceive it as an unpleasant itching sensation.

A great many nerve endings in the skin are responsible for the sensation of pain. Some are specialized to detect painful heat or cold, but most

Receptors located in the skin respond to a range of stimuli such as pressure, vibration and changes in temperature.

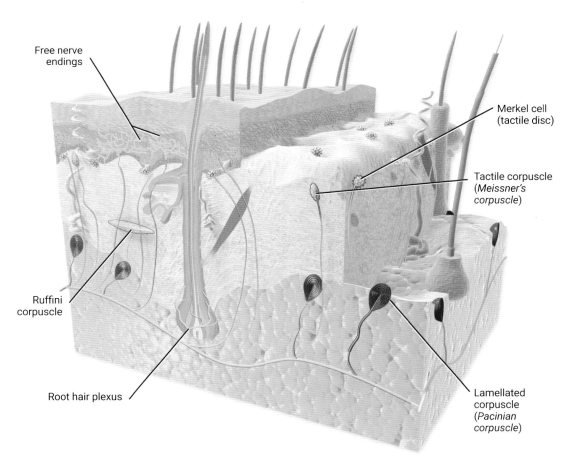

Free nerve endings

Merkel cell (tactile disc)

Tactile corpuscle (*Meissner's corpuscle*)

Ruffini corpuscle

Root hair plexus

Lamellated corpuscle (*Pacinian corpuscle*)

are responsive to multiple stimuli. A single nerve ending may respond to extreme pressure or to chemical signals from damaged cells as well as extremes of temperature, signalling to the brain that damage is being done.

PROPRIOCEPTION

Proprioception is the sense that tells us how we are oriented in space and where the parts of the body are located relative to each other. Part of proprioception is joint position sense, the ability to perceive the position of a joint without looking. It is what keeps us upright and balanced and able to walk or run without having to think about where to place each footstep.

The balance organs in the inner ear (see pages 127–8) are one component of a larger sensory system that contributes towards proprioception. It also takes in information from the eyes, and from receptors in the skin, muscles and joints that sense stretch, pressure and movement. All of this is processed in the brain, primarily in the cerebellum and brainstem, to provide an overview of how all of our body parts are positioned and moving through three-dimensional space.

As the brain takes in this information from the sensory systems about our position within the environment, it also sends out instructions to the muscular system to make any necessary adjustments. For example, information from the balance organs is used to control the muscles around the eyes, a process known as the vestibulo-ocular reflex, which moves the eyes to compensate for the movement of the body or the head.

Feedback from the proprioception system is what keeps us from falling over whenever we stand upright. Micro adjustments of the muscles keep us moving smoothly when we're walking or running over uneven ground or steady us when we have to stand on a moving train.

Kinaesthesia, a key part of hand-eye co-ordination and muscle memory, makes use of the proprioceptors in joints and muscles. It differs from proprioception in that it excludes the sense of equilibrium or balance. Kinaesthesia focuses on the body's movements and learning to adapt them, whereas proprioception is more concerned with awareness of those movements.

No sensory receptor cells are specifically specialized for proprioception. Muscle spindles, nested within and running parallel to muscle fibres, measure muscle stretch (muscle length) and transmit this information via associated neurons. Golgi tendon organs, located between muscle fibres and the tendons that attach them to bone, measure muscle tension. As well as their role in proprioception, muscle spindles play a part in fast, powerful muscle contractions and Golgi tendon organs help to prevent muscle injury due to overload.

The brain receives and interprets information from multiple inputs:

Vestibular organs in the inner ear send information about rotation, acceleration, and position

Eyes send visual information

Stretch receptors in skin, muscles and joints send information about the position of the body parts.

Proprioception, the awareness of the body's position in space, relies on inputs from a number of senses.

The Digestive System

The Digestive System

The digestive system has three main functions: breaking down food, absorbing its nutrients, and expelling any remaining waste. The components of the digestive system that accomplish these functions consist of the organs that make up the gastrointestinal (GI) tract, through which food actually passes, along with accessory organs which secrete enzymes and other substances necessary for digestion into the GI tract. The gastrointestinal tract is basically a long, continuous tube that extends from the mouth to the anus and includes the pharynx, oesophagus, stomach, and the small and large intestines. Fully extended, it is around 9 m (30 ft) long in adults and it takes up to 50 hours for food or food waste to travel all the way through.

The process of breaking down food into components the body can absorb is called digestion. Mechanical digestion is the physical breakdown of chunks of food into smaller pieces and takes place mainly in the mouth and stomach. Chemical digestion, the chemical breakdown of large, complex food molecules into smaller, simpler nutrient molecules that can be absorbed and utilized by the body, occurs mainly in the small intestine, though the process begins in the mouth and continues in the stomach.

After food is digested, the resulting nutrients are absorbed. Absorption, which also takes place principally in the small intestine, is the process by which substances pass into the bloodstream or lymph system to be circulated and supplied to the body's cells. Any remaining food material that has not been digested and absorbed is eliminated from the body through the anus.

THE MOUTH

The first organ of the GI tract, where food begins its journey into the body, is the mouth. This is where the mechanical digestion of food begins as the teeth crush and grind food into smaller pieces which mix with saliva, forming an easy to swallow soft lump called a bolus. The inside of the mouth is lined with mucous membrane, a mucus-producing tissue that also helps to moisten, soften and lubricate food. The mucous membrane is loosely connected to an underlying thin layer of smooth muscle, giving the mucous membrane considerable ability to stretch. The roof of the mouth, called the palate, separates the oral cavity from the nasal cavity. At the front of the palate the mucous membrane covers a plate of bone. The hard surface at the front of the palate allows pressure to be applied for chewing and mixing food. At the back, the mucous membrane overlies muscle and connective tissue and is softer and more pliable to accommodate the passage of food while swallowing. Muscles at either side of the soft palate contract to create the swallowing action.

Tongue

The fleshy, muscular, highly mobile tongue is attached to the floor of the mouth by a band of ligaments. The tongue can manipulate food for chewing and swallowing (and is also involved in speech). The upper surface of the tongue is covered with tiny papillae, which contain taste buds, the sensory detectors for the sense of taste (see page 130).

The organs of the digestive system.

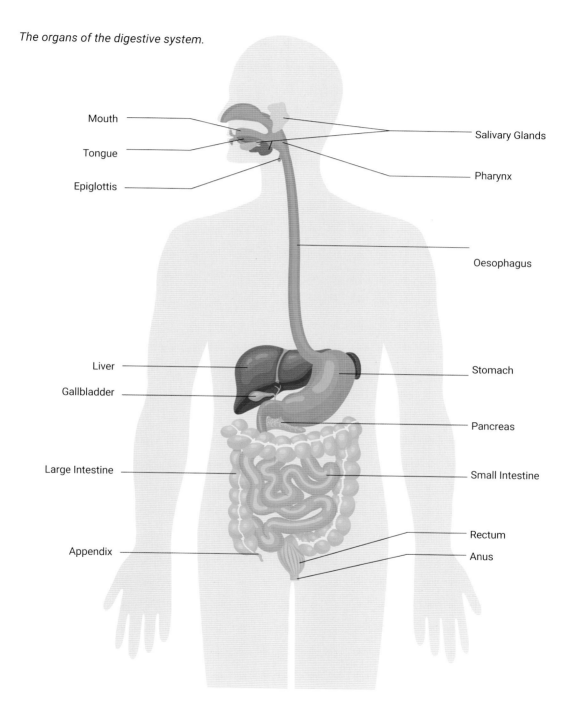

Mouth

Tongue

Epiglottis

Salivary Glands

Pharynx

Oesophagus

Liver

Gallbladder

Stomach

Pancreas

Large Intestine

Small Intestine

Appendix

Rectum

Anus

The mouth is the gateway to the digestive system where food is chewed and swallowed.

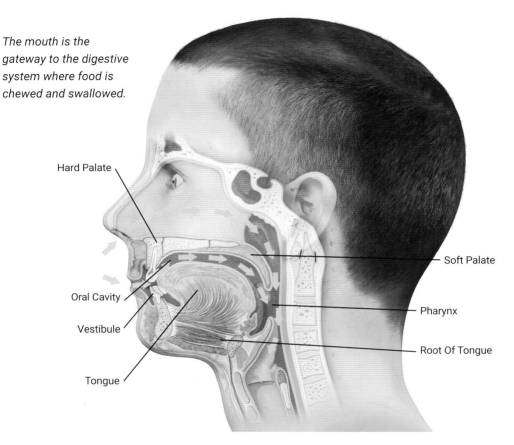

Hard Palate

Oral Cavity

Vestibule

Tongue

Soft Palate

Pharynx

Root Of Tongue

Teeth

The teeth are complex structures made of a bone-like material called dentin and covered with enamel, which is the hardest tissue in the body. Adults normally have a total of 32 teeth, with 16 in each jaw. The right and left sides of each jaw are mirror images of each other in terms of the teeth they contain. Teeth have different shapes suited for different aspects of chewing or mastication.

Incisors: sharp, blade-like teeth at the front of the mouth used for cutting or biting off pieces of food. Adults normally have four incisors in each jaw, making eight in total.

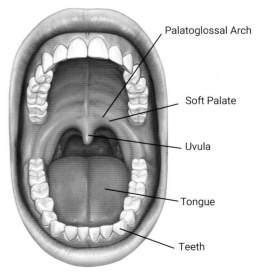

Palatoglossal Arch

Soft Palate

Uvula

Tongue

Teeth

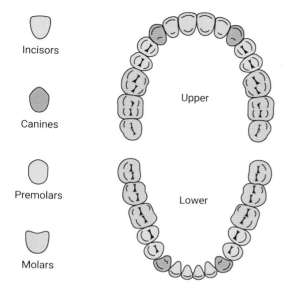

Incisors

Canines

Premolars

Molars

Upper

Lower

Canines: pointed teeth on either side of the incisors used for tearing tough or stringy food. Adults normally have two canines in each jaw, making four in total.

Premolars and *molars:* cuboid teeth with cusps and grooves located on the sides and towards the back of the jaws. Premolars are closer to the front of the mouth, molars at the rear. Molars are larger and have more cusps but both are used for crushing and grinding food. Adults normally have two premolars and three molars on each side of each jaw, making eight premolars and twelve molars in total.

Salivary glands

The sight, smell and taste of food stimulates the release of digestive enzymes and other secretions

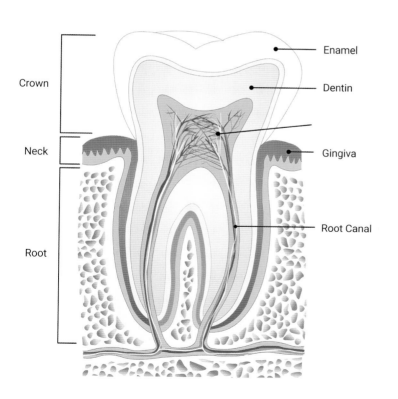

Crown

Neck

Root

Enamel

Dentin

Gingiva

Root Canal

Above left: An adult human has 32 permanent teeth, 16 in each jaw.

Left: A cross-section through a tooth.

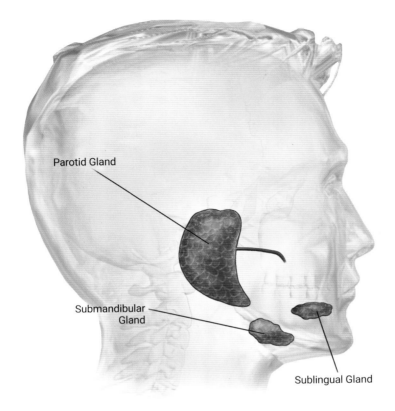

Parotid Gland

Submandibular Gland

Sublingual Gland

The location of the major salivary glands around the mouth.

by salivary glands inside the mouth. There are three pairs of major salivary glands, all exocrine glands, which secrete saliva into the mouth through ducts. Saliva mainly aids the digestive process, but also contains antibodies that guard against infection and helps to maintain dental health by cleaning the teeth. By keeping the mouth lubricated, saliva also facilitates the mouth movements of speech. The largest of the major salivary glands are the parotid glands, located on either side of the mouth in front of the ears. The next largest pair are the submandibular glands, found beneath the lower jaw. The third pair, the sublingual glands, are located beneath the tongue.

In addition to the major salivary glands, hundreds of minor salivary glands are found in the tongue and lining of the mouth. The major glands and most of the minor glands secrete amylase, a digestive enzyme that begins the chemical digestion of starch. The minor salivary glands on the tongue also secrete the fat-digesting enzyme lipase.

THE PHARYNX

The tube-like pharynx is the gateway to both the digestive system and the respiratory system. As part of the respiratory system, it provides a passage for air to travel from the nasal cavity to the larynx. Its role in the digestive system is to allow swallowed food to pass from the oral cavity into the oesophagus. During swallowing, the tongue moves backwards causing a flap of elastic cartilage called the epiglottis to close over the opening to the larynx, preventing food or drink from entering it. A process called peristalsis (see below) starts at the top of the oesophagus when food is swallowed and continues down the oesophagus in a single wave, pushing the bolus of food ahead of it.

THE OESOPHAGUS

The oesophagus is a muscular tube through which food travels from the pharynx to the stomach. In adults, the oesophagus is around 25 cm (10 in) in length, the actual length depending on the person's height. When food is not being swallowed, the oesophagus is closed at either end by the upper and lower oesophageal sphincters, rings of muscle that contract to close off openings. The act of swallowing triggers the opening of

Peristalsis involves a wave-like contraction of smooth muscles that pushes food on its way through the digestive tract.

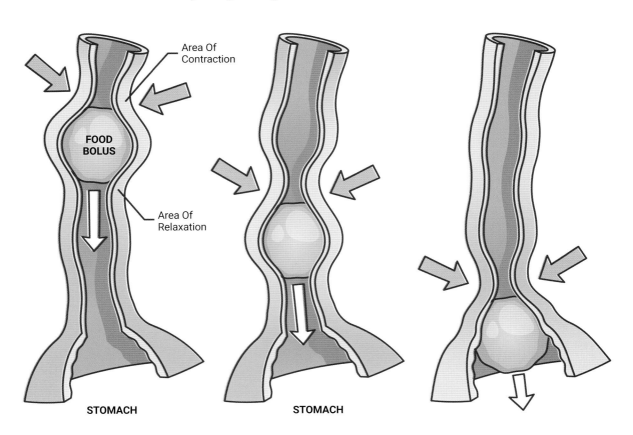

Area Of Contraction

FOOD BOLUS

Area Of Relaxation

STOMACH

STOMACH

the upper sphincter, allowing the food bolus to enter the oesophagus from the pharynx. The sphincter then closes again to prevent the bolus from moving back into the pharynx.

The food bolus is pushed through to the stomach by a series of strong wave-like contractions of the smooth muscle lining the oesophagus. These contractions are called peristalsis and will propel your food on its way even if you are standing on your head. The inner lining of the oesophagus is mucous membrane, which provides a smooth, slippery surface to

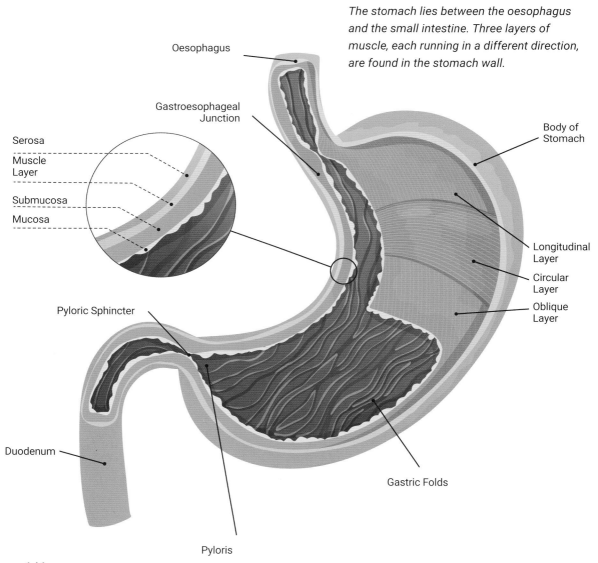

The stomach lies between the oesophagus and the small intestine. Three layers of muscle, each running in a different direction, are found in the stomach wall.

Oesophagus

Gastroesophageal Junction

Serosa

Muscle Layer

Submucosa

Mucosa

Body of Stomach

Longitudinal Layer

Circular Layer

Oblique Layer

Pyloric Sphincter

Gastric Folds

Duodenum

Pyloris

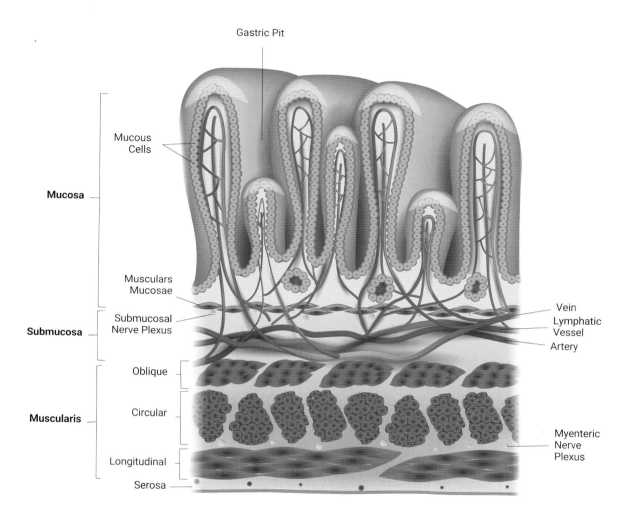

Gastric Pit

Mucous Cells

Mucosa

Musculars Mucosae

Submucosal Nerve Plexus

Submucosa

Oblique

Circular

Muscularis

Longitudinal

Serosa

Vein

Lymphatic Vessel

Artery

Myenteric Nerve Plexus

ease the passage of food. The membrane cells are constantly replaced as the food passing over them wears them away. The oesophagus passes through an opening in the diaphragm, the sheet of muscle that separates the abdomen from the thorax, before reaching the stomach. The lower oesophageal sphincter, located at the junction between the oesophagus and the stomach, opens when the food bolus reaches it, allowing the food to enter the stomach. Normally this sphincter

Mucous cells in the lining of the stomach secrete a protective coat of alkaline mucus, protecting the lining from the highly acid gastric juice produced in the gastric pits. Beneath this, the muscular layers mix and churn the stomach contents.

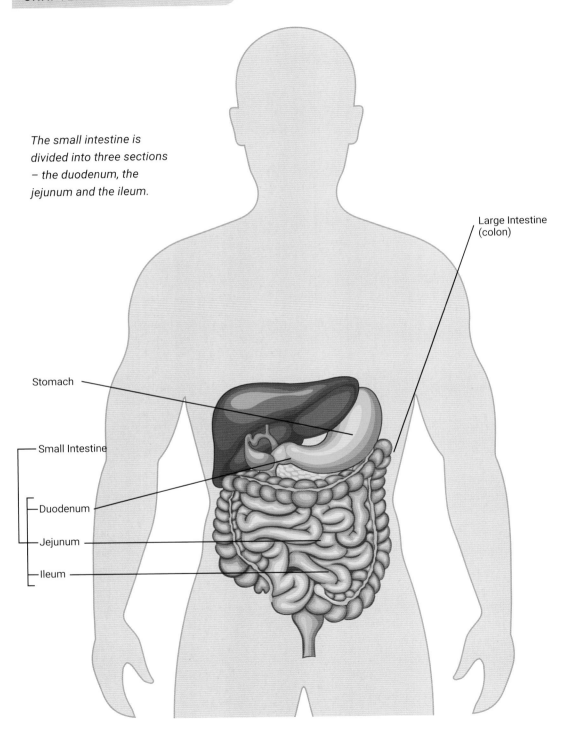

The small intestine is divided into three sections – the duodenum, the jejunum and the ileum.

Large Intestine (colon)

Stomach

Small Intestine

Duodenum

Jejunum

Ileum

remains closed to prevent stomach acid from entering the oesophagus. If it fails to do so it can lead to painful heartburn.

THE STOMACH

The stomach is a J-shaped organ joined at its upper end to the oesophagus and at its lower end to the duodenum, the first part of the small intestine. Empty, the stomach has a volume of around 75 millilitres, but can expand to hold up to about 1 litre (35 fluid oz) of food. Waves of peristalsis passing through the muscular wall of the stomach mix and churn the food inside. These movements continue the process of mechanical digestion that began by chewing in the mouth. The churning also mixes the food with stomach secretions, including gastric acid (mainly hydrochloric acid) that aid in its chemical digestion.

The highly acidic conditions inside the stomach are necessary for the effective functioning of the enzyme pepsin (also secreted by the stomach) which begins the digestion of protein. The stomach is protected from the gastric acid by a coating of mucus secreted by the stomach lining. After about an hour the food in the stomach has been reduced to a thick, semi-liquid called chyme. The pyloric sphincter between the stomach and duodenum opens to allow the chyme to enter the small intestine, where the process of digestion and absorption continues.

THE SMALL INTESTINE

The small intestine lies between the stomach and large intestine. It is termed 'small' because it has a smaller diameter than the large intestine (approximately 2.5 to 3 cm (1 in) as opposed to around 7 cm (2.75 in)). Its average length in an adult is around 3 m (10 ft) (less in females than in males). The internal surface area of the small intestine totals an average of about 30 m² (320 ft²), almost half the size of a badminton court.

The surface area is so large because the mucous membrane, or mucosa, lining the small intestine is highly folded in such a way that the chyme spirals through rather than travelling in a straight line, meaning that it moves through the intestine more slowly. These folds are covered with tiny finger-like projections called villi (singular: villus) about 0.5 to 1 mm long. There are between 20 and 40 villi per square millimetre of the mucosa, increasing its surface area hugely. In addition, the individual cells on the surface of the villi are themselves covered with projections, called microvilli. There are an estimated 200 million microvilli in every square millimetre of the intestinal membrane. Altogether, folds, villi and microvilli constitute a tremendous surface area over which chyme can come into contact with digestive enzymes, and through which nutrients can be absorbed. Inside each villus, a network of tiny blood and lymph vessels take the absorbed nutrients which will eventually be transported to all parts of the body. Nutrient-rich blood from the small intestine is carried to the liver via the hepatic portal vein.

The Duodenum

Just 25 cm (10 in) in length in the average adult, the C-shaped duodenum curving around the pancreas, is the first and shortest part of the small intestine. This is where most of the chemical digestion of the digestive system takes place. Carbohydrates, lipids and proteins are all digested here.

The partially digested, semi-liquid chyme that enters the duodenum from the stomach is joined by digestive enzymes and alkaline bicarbonate

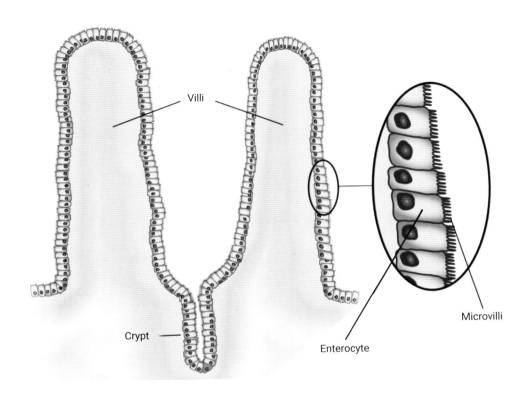

Villi

Microvilli

Crypt

Enterocyte

The finger-like villi and hair-like microvilli in the lining of the small intestine greatly increase the surface area over which nutrients can be absorbed.

from the pancreas via the pancreatic duct, and bile from the liver via the gallbladder and the common bile duct. The pancreas secretes up to 1.5 litres (50 fluid oz) of pancreatic juice, mostly consisting of water, into the duodenum every day. The pancreas and gallbladder empty into the duodenum at a feature called the ampulla of Vater.

The lining of the duodenum also secretes digestive enzymes and contains glands, called Brunner's glands, that secrete mucus and bicarbonate. The bicarbonate from the pancreas and Brunner's glands, along with bile from the

liver, serve to neutralize the highly acidic chyme from the stomach. This is necessary because the digestive enzymes in the duodenum, unlike the pepsin in the stomach, require a nearly neutral environment in order to work effectively.

Complex carbohydrates such as starch are broken down by the digestive enzyme amylase, produced by the pancreas, and by duodenal enzymes, such as sucrase which breaks down sucrose, and lactase which breaks down lactose. The enzymes trypsin and chymotrypsin produced in the pancreas break down proteins into smaller peptides which are in turn broken down into amino acids by pancreatic enzymes called peptidases. Triglycerides are broken down into fatty acids and glycerol by lipase from the pancreas which works with bile secreted by the liver and stored in the gallbladder. Bile acts in a similar way to the detergent you might use to clean a frying pan, surrounding fats and forming an emulsion of smaller particles called micelles which disperse into the watery environment of the duodenum, making them more accessible to the lipase.

The Jejunum

The middle part of the small intestine, following the duodenum, is called the jejunum. The jejunum is about 1 m (3 ft) long. Its main function is absorption of the products of digestion. All nutrients are absorbed into the blood, with the exception of fatty acids and fat-soluble vitamins, which are absorbed into the lymph.

The Ileum

The longest part of the small intestine is the ileum, measuring about 1.8 m (6 ft) in length.

The pancreas releases acid-neutralizing bicarbonate and digestive enzymes into the duodenum through the pancreatic ducts.

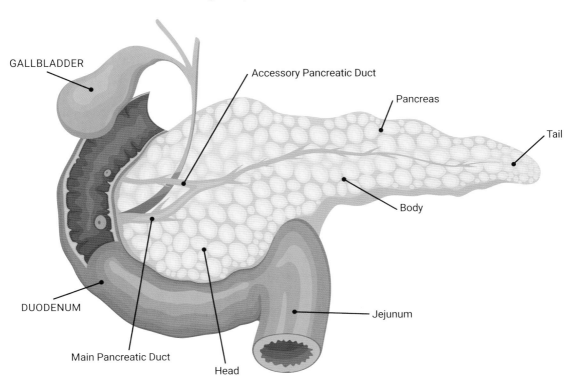

GALLBLADDER

Accessory Pancreatic Duct

Pancreas

Tail

Body

DUODENUM

Jejunum

Main Pancreatic Duct

Head

It is thicker, has more developed mucosal folds than the jejunum and is better supplied with blood and lymph vessels. The main function of the ileum is to absorb vitamin B12 and bile salts and any other remaining nutrients that were not absorbed in the jejunum. Some cells in the lining of the ileum produce enzymes that catalyse the final stages of digestion of any undigested protein and carbohydrate molecules. All substances that remain undigested or unabsorbed by the time they reach the end of the ileum pass into the cecum, the first part of the large intestine, at the ileocecal sphincter.

THE LARGE INTESTINE

The large intestine is the final part of the gastrointestinal tract. It is about half the length of the small intestine but more than twice its diameter and surrounds the small intestine on three sides. Its functions are to complete the absorption of nutrients and water, to synthesize certain vitamins, and to form faeces, and eliminate them from the body. Salts, such as sodium, are also removed for recycling before the wastes are eliminated from the body. The chyme that enters the large intestine contains few nutrients except water (90 per cent of which is absorbed by the small intestine), and, if made necessary by injury or illness, the large intestine can be completely removed without significantly affecting digestive functioning.

The large intestine is also home, particularly in the descending colon, to huge numbers of beneficial bacteria which ferment many of the undigested and unabsorbed materials in food waste. The waste products of this bacterial breakdown are nitrogen, carbon dioxide, methane and other gases responsible for intestinal gas, or flatulence. Some bacteria also produce vitamins, including B1 (thiamine), B2 (riboflavin), B7 (biotin), B12, and K, that are absorbed from the colon. The bacteria in the colon may inhibit the growth of harmful bacteria and also stimulate the immune system to produce antibodies that are effective against these pathogenic bacteria, thereby helping to prevent infections. Other roles played by bacteria in the large intestine include breaking down toxins.

The Cecum

The cecum is a sac-like structure about 6 cm (2.4 in) long that is suspended beneath the ileocecal valve. It receives the contents of the ileum and continues the absorption of water and salts. The 7.6 cm (3 in) long appendix (or vermiform appendix) is a winding tube attached to the cecum. Although the appendix contains tissue that suggests it was part of the immune system, it is now generally considered vestigial, meaning that its original function has been lost. Appendicitis, the inflammation of the appendix, is a fairly common medical problem, easily resolved by surgically removing the offending organ. People who have had this procedure carried out do not appear to suffer any ill effects, which appears to indicate that the appendix is indeed dispensable.

The Colon

Food residue leaving the cecum travels up into the ascending colon on the right-hand side of the abdomen. At the surface of the liver, the colon bends to form the right colic flexure and becomes the transverse colon. The food residue continues through the transverse colon, travelling across to the left side of the abdomen, where the colon angles sharply at the spleen, at the left colic flexure. It then goes into the descending colon, which runs down the left side of the abdominal

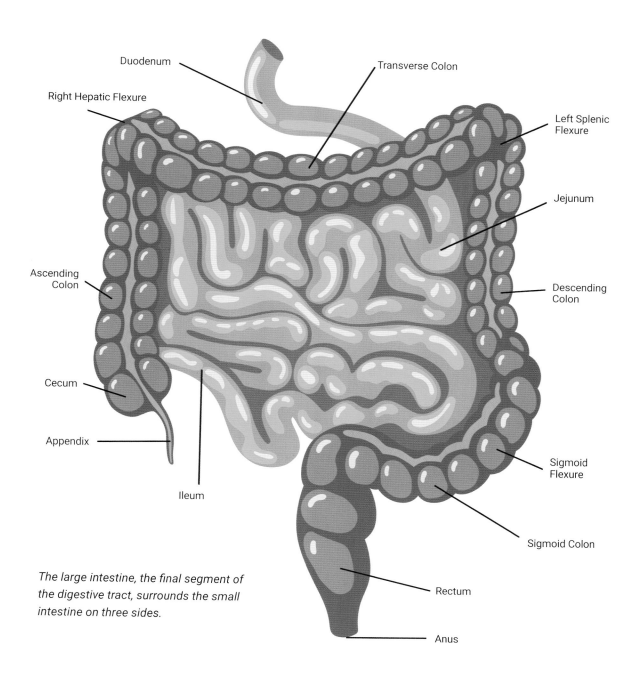

Duodenum

Transverse Colon

Right Hepatic Flexure

Left Splenic Flexure

Jejunum

Ascending Colon

Descending Colon

Cecum

Appendix

Ileum

Sigmoid Flexure

Sigmoid Colon

Rectum

Anus

The large intestine, the final segment of the digestive tract, surrounds the small intestine on three sides.

ANTERIOR VIEW

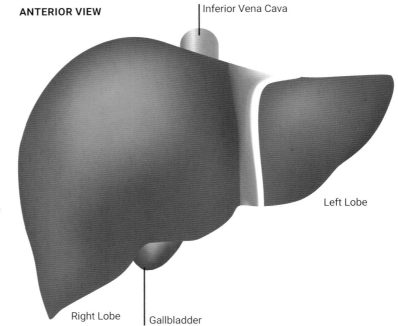

Inferior Vena Cava

The liver is one of the body's vital organs. Blood from the digestive system enters the liver through the portal vein so that waste products and toxins can be broken down.

Left Lobe

Right Lobe

Gallbladder

POSTERIOR VIEW

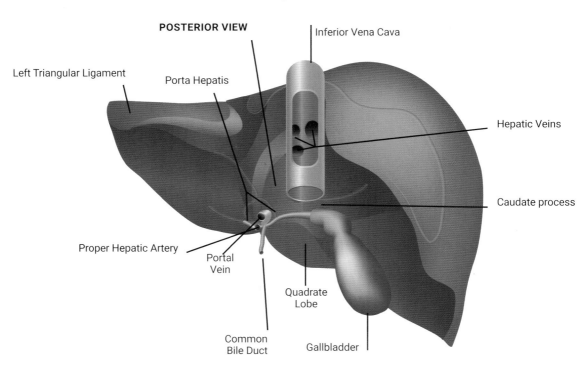

Inferior Vena Cava

Left Triangular Ligament

Porta Hepatis

Hepatic Veins

Caudate process

Proper Hepatic Artery

Portal Vein

Quadrate Lobe

Common Bile Duct

Gallbladder

wall. After entering the pelvis, it becomes the S-shaped sigmoid colon.

THE RECTUM

Food residue leaves the sigmoid colon and enters the rectum, the final 20 cm (8 in) or so of the digestive tract. Located in the pelvis, the rectum follows the curved contour of the sacrum and has a trio of internal transverse folds called the rectal valves. These valves help separate the faeces from gas to prevent the simultaneous passage of faeces and gas. Which is perhaps something best avoided.

Faeces is composed of undigested food residues, substances that were digested but not absorbed, millions of bacteria, old epithelial cells from the digestive tract's mucous membranes, inorganic salts, and enough water to let it pass smoothly out of the body. Of every 500 millilitres (17 fluid oz) of food residue that enters the cecum each day, about 150 millilitres (5 fluid oz) become faeces. The rectum stores faeces until elimination occurs when faeces passes through the final part of the large intestine, called the anus, an opening to the outside of the body. The exit of faeces from the body is regulated by two sphincters; the inner sphincter is involuntary and the outer sphincter is voluntary.

ACCESSORY DIGESTIVE ORGANS

Chemical digestion in the small intestine relies on the activities of three accessory digestive organs, which, although not part of the gastrointestinal tract, are vital to its functioning. These are the liver, pancreas and gallbladder. The digestive role of the liver is to produce bile which is stored and concentrated in the gallbladder for release into the duodenum. The pancreas produces pancreatic fluid containing digestive enzymes and bicarbonate, which is also delivered to the duodenum.

The Liver

The liver is the largest gland in the body. It is divided into two primary lobes: a large right lobe and a much smaller left lobe. Weighing about 1.5 kg (3.3 lb) in an adult, it is the heaviest of the body's internal organs. The liver lies beneath the diaphragm in the right upper quadrant of the abdominal cavity where it is protected by the surrounding ribs. The liver has two main sources of blood. The hepatic artery and the hepatic portal vein enter the liver at the porta hepatis – the gateway to the liver. All blood from the gastrointestinal tract passes through the liver. The liver may hold between 10 and 13 per cent of the body's blood supply at any one time.

The hepatic portal vein delivers blood containing nutrients absorbed from the small intestine and actually supplies more oxygen to the liver than the much smaller hepatic artery. In addition to nutrients, drugs and toxins are also absorbed. After processing the nutrients and toxins, the liver releases nutrients needed by other cells back into the blood, via the hepatic vein and the inferior vena cava.

The main digestive function of the liver is the production of bile, a yellowish alkaline liquid that consists of a mixture of water, electrolytes, bile salts, cholesterol and other substances. Some of the components of bile are synthesized by liver cells called hepatocytes, while others are extracted from the blood. Bile is secreted into small ducts that unite to form the larger right and left hepatic ducts, which themselves merge and exit the liver as the common hepatic duct. This duct then joins with the cystic duct from the gallbladder, to form the common bile duct

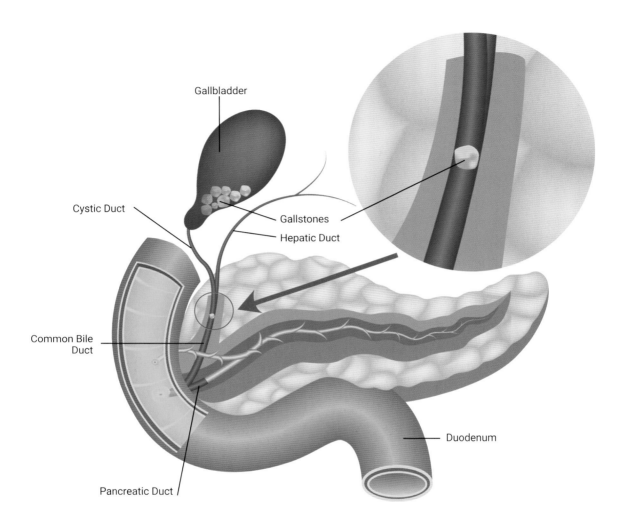

Gallbladder

Cystic Duct

Gallstones

Hepatic Duct

Common Bile
Duct

Duodenum

Pancreatic Duct

Excess cholesterol in bile may lead to the formation of gallstones. If these become trapped in the bile duct they can cause great pain.

through which bile flows into the duodenum. In the duodenum, the bile neutralizes acidic chyme from the stomach and emulsifies fat globules into smaller particles that are easier to digest chemically. Bile also aids with the absorption of vitamin K. Bile that is secreted when digestion is not taking place is stored in the gallbladder, which also feeds into the duodenum via the common bile duct.

The Gallbladder

The gallbladder is a small, pouch-like organ about 8 to 10 cm (3–4 in) long located just beneath the right side of the liver. Bile leaving the liver reaches the gallbladder through the cystic duct. The gallbladder stores and concentrates between 30 to 60 millilitres of bile until it is needed in the duodenum. The presence of fat in the duodenum stimulates the release of a hormone which signals the gallbladder to contract and force its contents back through the cystic duct and into the common bile duct to drain into the duodenum.

If the levels of cholesterol in bile stored inside the gallbladder become too high the excess cholesterol forms into stones. Gallstones are very common. More than 1 in every 10 adults in the UK are likely to have gallstones, although only a minority will develop symptoms, which can include intense pain if the stone gets trapped in a duct. The worst cases are treated by surgically removing the gallbladder altogether.

Along with its role in digestion, the liver has a number of other important functions to perform and is one of the most vital organs in the human body. The liver plays a part in regulating blood sugar levels, synthesizing glycogen from glucose and storing it until it is required, when it then breaks down the stored glycogen back into glucose and releases it back into the bloodstream. The liver is also a repository for many other substances, including vitamins A, D, B12, E and K and the minerals iron, zinc and copper.

Numerous proteins and their amino acid building blocks are synthesized in the liver, including fibrinogen, which is involved in blood clotting; insulin-like growth factor (IGF-1), which is important for childhood growth; and albumin, the most abundant protein in blood serum, which transports fatty acids and steroid hormones in the blood. Many important lipids, including cholesterol, triglycerides and lipoproteins are also synthesized in the liver.

Waste products and toxic substances are broken down by the liver. The breakdown products are excreted in bile or travel to the kidneys, to be excreted in urine. The liver also helps to purify the blood. Cells in the liver called Kupffer cells are phagocytes, consuming dead blood cells as well as bacteria in the blood.

The Pancreas

The pancreas is part of both the digestive system and the endocrine system (see Chapter Thirteen). It is about 15 cm (6 in) long and located in the abdomen behind the stomach, with the head of the pancreas surrounded by the duodenum. It has two major ducts: the main pancreatic duct and the accessory pancreatic duct, both of which drain into the duodenum.

In its role as part of the digestive system, the pancreas secretes bicarbonate, which helps neutralize acidic chyme from the stomach, along with a number of digestive enzymes. The pancreatic digestive enzymes are secreted by clusters of cells called acini, and travel through the pancreatic ducts to the duodenum. The pancreas is stimulated to produce these substances by hormones that are released into the bloodstream when food is detected in the stomach and duodenum. The pancreatic digestive enzymes include amylase, which breaks down carbohydrates; trypsin and chymotrypsin, which break down proteins; lipase, which breaks down lipids; and deoxyribonucleases and ribonucleases, which break down nucleic acids.

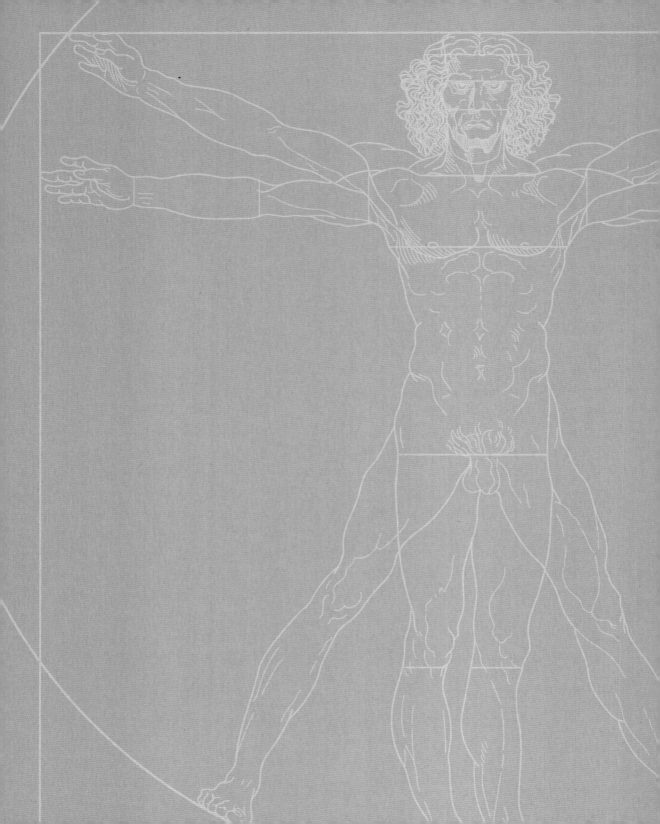

CHAPTER NINE

The Renal System

The Renal System

The renal system, also known as the urinary system, consists of the kidneys, ureters, bladder and urethra. Its main function is the elimination of the waste products of metabolism from the body, which it does by forming and excreting urine.

Excretion is the process by which wastes and excess water are removed from the body. It is an essential process for the maintenance of homeostasis (see pages 28–31) in the body and helps to prevent damage being done by the build-up of toxic wastes. Some of the waste products that must be excreted from the body include carbon dioxide from cellular respiration, ammonia and urea from the breakdown of proteins, and uric acid from the breakdown of nucleic acids.

Along with the organs of the renal system other organs of the body involved in waste disposal include the skin, liver, large intestine and lungs. Although they could be said to make up an excretory system, they don't work together in the same way that the organs of the digestive system or the circulatory system do, for example.

The skin is part of the integumentary system (see pages 44–51), but through the production of sweat it also plays a part in eliminating excess water and salts, as well as a small amount of urea. The liver has a number of major roles in the body, just one of which is as an organ of excretion. For example, it transforms ammonia, a poisonous by-product of the breakdown of proteins, into urea, which is filtered from the blood by the kidneys and excreted in urine. The main function of the large intestine, part of the digestive system, is to eliminate solid wastes that remain after the digestion of food and to extract water from indigestible matter in food waste. The large intestine also collects wastes such as the pigment bilirubin from the liver. Bilirubin, one of the wastes produced from the breakdown of old red blood cells, gives human faeces its characteristic brown colour.

The lungs, part of the respiratory system, are responsible for the excretion of gaseous wastes from the body, mainly carbon dioxide, a waste product of cellular respiration.

A healthy adult typically produces between 1 and 2 litres (34–68 fluid oz) of urine every day. The main waste products removed in urine are urea, a by-product of protein breakdown, and uric acid, a by-product of nucleic acid breakdown. Excess water and mineral ions are also eliminated in urine.

In addition to the elimination of waste products, the urinary system fulfils several other important functions. These include maintaining blood pH through the excretion of bicarbonate. When pH is too low (too acidic), the kidneys excrete less bicarbonate (which is alkaline) in urine; when it is too high (too alkaline), more bicarbonate is excreted.

The formation of urine is closely regulated by several endocrine hormones, including antidiuretic hormone, parathyroid hormone and aldosterone. One of the main functions of antidiuretic hormone (ADH), is conserving body water. It is released by the posterior pituitary gland when the body is dehydrated, causing the kidneys to excrete less water in urine. Parathyroid hormone regulates the balance of

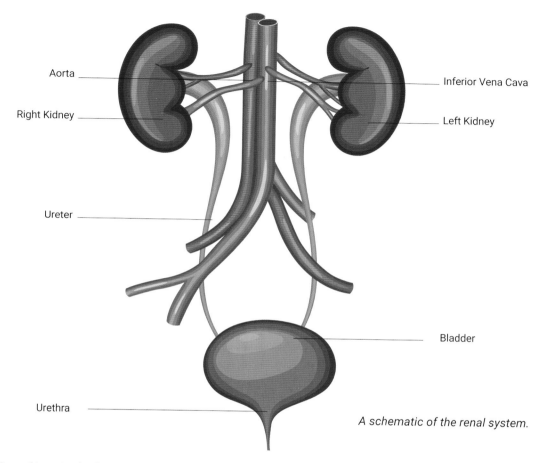

Aorta

Right Kidney

Ureter

Urethra

Inferior Vena Cava

Left Kidney

Bladder

A schematic of the renal system.

mineral ions in the body, stimulating the kidneys to excrete less calcium and more phosphorus in urine. Aldosterone plays a role in regulating blood pressure. It is secreted by the cortex of the adrenal glands above the kidneys, causing them to excrete less sodium and water in urine.

THE KIDNEYS

The two bean-shaped kidneys (kidney beans are named for their resemblance to the kidney rather than the other way around) are located just below the diaphragm, high in the back of the abdominal cavity, one on each side of the spine. Each kidney is about 11 cm (4 in) long and weighs around 150 grams (5 oz). The entire kidney is surrounded by tough, fibrous tissue, called the renal capsule, and by two layers of protective fat. The right kidney is slightly smaller and lower and sits behind the liver; the slightly larger left kidney sits behind the spleen. Each kidney has a convex (curving outwards) side and a concave (curving inwards) side.

Blood is supplied to the kidneys via the left (to the left kidney) and right (to the right kidney) renal arteries. The renal arteries branch directly from the aorta, the largest artery in the body. The

The internal structure of a kidney.

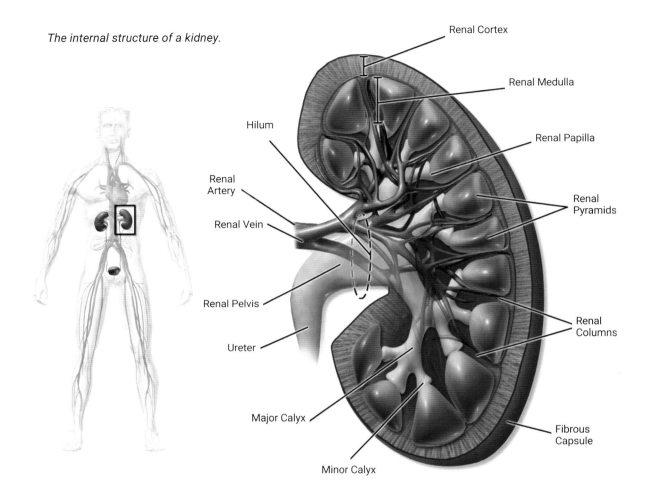

Renal Cortex

Renal Medulla

Renal Papilla

Hilum

Renal Artery

Renal Vein

Renal Pyramids

Renal Pelvis

Renal Columns

Ureter

Major Calyx

Fibrous Capsule

Minor Calyx

renal artery enters the kidney on the concave side in an area of the kidney called the hilum. This is also where the renal vein and ureter leave the kidney.

Each kidney is divided into two major layers internally: the outer renal cortex and the inner renal medulla. These layers are formed from many cone-shaped renal lobules, each of which is made up of renal cortex surrounding a portion of the medulla called a renal pyramid. Located within the renal pyramids are the nephrons, the structural and functional units of the kidney. Projections of the cortex called renal columns protrude between the renal pyramids. The tip of each pyramid empties urine into a chamber called a minor calyx. Several minor calyces empty into a major calyx, which in turn empties into the funnel-shaped renal pelvis, which becomes the ureter as it leaves the kidney.

The renal arteries carry waste products in the blood into the kidneys. Within the kidney, the renal artery branches into increasingly smaller arteries and then into arterioles that penetrate the renal pyramids and feed into the nephrons, the structures in the kidney that are responsible for filtering the blood. After the filtered blood has passed through the nephrons it moves into a network of venules that converge to form small and then increasingly larger veins that merge into the renal vein, which carries the filtered blood away from the kidney to the inferior vena cava.

The Nephron

Each kidney typically contains around a million nephrons, the functional units of the kidney. Each nephron acts as an independent filter and urine-processing unit, cleansing the blood of toxins and maintaining a homeostatic balance through the processes of filtration, reabsorption and secretion.

Each nephron has a specialized network of ducts called a renal tubule along with a renal corpuscle which consists of a network of capillaries held inside a cup-shaped structure called the Bowman's, or glomerular, capsule, from which the tubule extends. The end of the tubule nearest the Bowman's capsule is called the proximal convoluted tubule. The tubule continues as a loop called the loop of Henle, which becomes the distal convoluted tubule and finally joins with a collecting duct. Arterioles surround the total length of the renal tubule in a mesh called the peritubular capillary network.

Blood enters the nephron through the afferent arteriole. Some of the blood passes through the capillaries of the glomerulus and some continues on through the efferent arteriole, which follows the renal tubule of the nephron. Blood from the afferent arteriole is under pressure as it flows through the glomerular capillaries, resulting in water and solutes being filtered out of the blood and into the Bowman's capsule. This is the filtration stage of nephron function. The filtered substances, called filtrate, pass from the Bowman's capsule into the proximal end of the renal tubule. The filtrate includes water, salts, nutrients and waste products such as urea. Anything too large to move through the pores in the glomerulus, such as proteins and blood cells, stays in the capillaries.

As filtrate moves through the renal tubule, some of the substances it contains are reabsorbed from the filtrate back through the peritubular capillary network and into the blood in the efferent arteriole. This is the reabsorption stage of nephron function which occurs mainly in the proximal convoluted tubule and the loop of Henle. It ensures that useful materials are returned to the bloodstream rather than being excreted in urine and lost to the body. Most of the glucose and amino acids previously filtered out are reabsorbed in the peritubular capillary network along with about two-thirds of the filtered salts and water.

Additional reabsorption takes place at the distal end of the renal tubule. Other substances from the blood are added to the filtrate in the tubule at this point in the secretion stage of nephron function. Reabsorption and secretion in the distal convoluted tubule are largely under the control of endocrine hormones that maintain homeostasis of water and mineral salts in the blood by controlling what is reabsorbed from the filtrate and what is secreted into the filtrate to become urine.

Urine empties from the distal end of the renal tubule into increasingly larger collecting ducts. More water may be reabsorbed as the urine passes through the collecting ducts, which is a process controlled by antidiuretic hormone

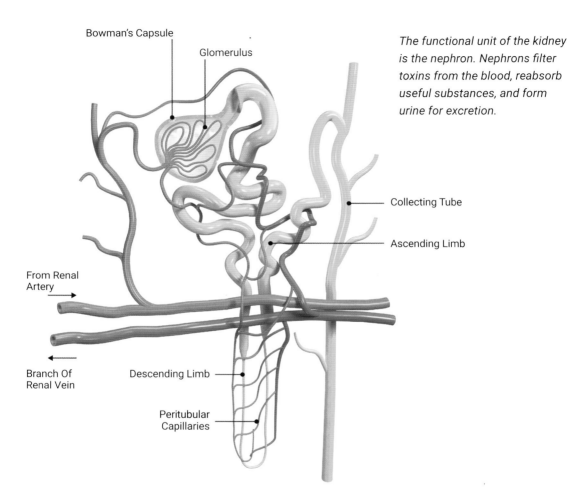

Bowman's Capsule

Glomerulus

The functional unit of the kidney is the nephron. Nephrons filter toxins from the blood, reabsorb useful substances, and form urine for excretion.

Collecting Tube

Ascending Limb

From Renal Artery

Branch Of Renal Vein

Descending Limb

Peritubular Capillaries

from the posterior pituitary gland. This hormone makes the collecting ducts permeable to water, allowing water molecules to pass through them into capillaries by osmosis. As much as 75 per cent of the water may be reabsorbed, making the urine more concentrated.

Urine finally exits the largest collecting ducts through the renal papillae, emptying into the renal calyces, and finally into the renal pelvis. From there, it travels through the ureter to the urinary bladder for eventual excretion from the body. On average, an adult will excrete roughly 1.5 litres (50 fluid oz) of urine a day.

URETERS AND BLADDER

The tube-like ureters, one for each kidney, connect the kidneys with the urinary bladder. In adults, ureters are between 25 and 30 cm (10–12 in) long and around 3 to 4 mm in diameter. Each ureter arises in the pelvis of a kidney, passing down the side of the kidney, and entering the back of the bladder. Sphincters on the ureters where they enter the bladder prevent the urine from flowing back to the kidney.

The outermost layer of the ureter walls is formed of fibrous tissue. Inside this, two layers of smooth muscle, an inner circular layer and an outer longitudinal layer, around the ureter, can contract in peristaltic waves to propel urine from the kidneys to the urinary bladder. The next layer of the ureter walls is made up of loose connective tissue that contains elastic fibres, nerves, and blood and lymphatic vessels. The innermost layer of the ureter wall is lined with transitional epithelium. Unlike other types of epithelium, transitional epithelium is capable of stretching and does not produce mucus. It also lines much

of the urinary system, including the renal pelvis, bladder and much of the urethra, allowing these organs to stretch and expand as they fill with urine or as urine passes through them.

The muscular and stretchy urinary bladder is a hollow organ that rests on the pelvic floor. Urine from the kidneys enters the urinary bladder from the ureters through two openings on either side of the back wall of the bladder. It is collected and stored here before being eliminated through urination. Urine leaves the bladder through a sphincter called the internal urethral sphincter, which relaxes and opens to allow urine to flow out of the bladder and into the urethra.

The outer covering of the bladder is formed of peritoneum, a smooth layer of epithelial cells that lines the abdominal cavity and covers most

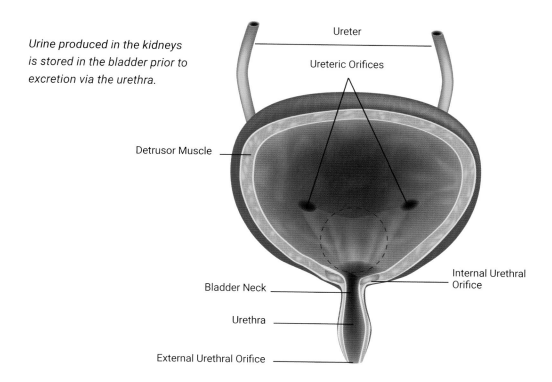

Urine produced in the kidneys is stored in the bladder prior to excretion via the urethra.

Ureter

Ureteric Orifices

Detrusor Muscle

Internal Urethral Orifice

Bladder Neck

Urethra

External Urethral Orifice

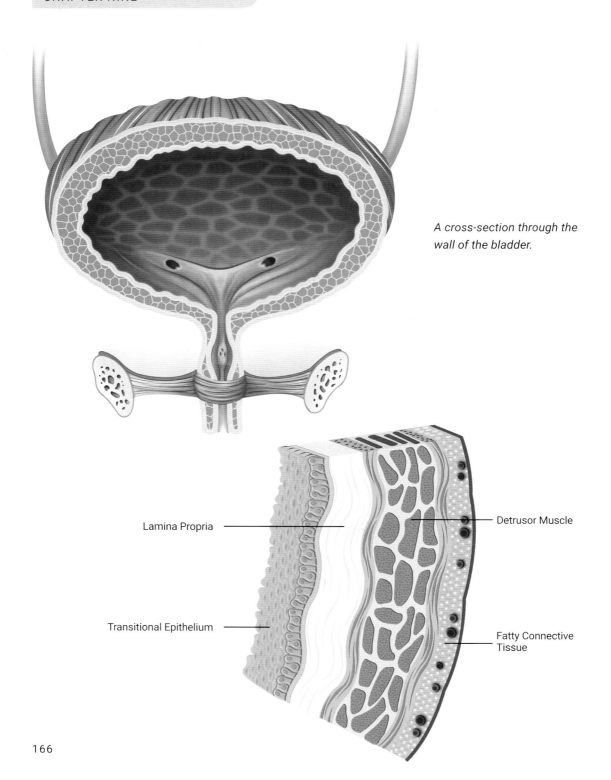

A cross-section through the wall of the bladder.

Lamina Propria

Transitional Epithelium

Detrusor Muscle

Fatty Connective Tissue

abdominal organs. As already mentioned, the bladder is lined with transitional epithelium, which can stretch as the bladder fills with urine. The next layer, called the lamina propria, is formed from loose connective tissue, supplied with nerves, and blood and lymphatic vessels. Next comes a submucosa layer, connecting the lining of the bladder with the detrusor muscle, a smooth muscle in the walls of the bladder. The detrusor muscle is controlled by both the autonomic and somatic nervous systems. It relaxes to allow the bladder to hold more urine, but when the bladder is about half full the stretching of its walls signals the need to urinate. Nervous signals under conscious control cause the detrusor muscle to contract and the internal urethral sphincter to relax and open, with the result that urine is expelled from the bladder and into the urethra.

THE URETHRA

This connects the urinary bladder to the external urethral orifice, where the urethra opens to the outside world. The urethra in males averages about 20 cm (8 in) whereas in females it is only about 5 cm (2 in). This difference in length is accounted for by the necessity of the urethra to travel through the male penis. In males, the urethra carries semen as well as urine, but in females it carries only urine. The urethra additionally passes through the male prostate gland, which is absent in women.

The first two-thirds of the urethra (leaving the bladder) are lined with transitional epithelium, and the final third with mucus-secreting epithelium which helps protect the epithelium from the corrosive effects of urine. Layers of smooth muscle around the urethra relax to allow the urine to pass through when the bladder contracts to force it out, and the external urethral sphincter relaxes and opens the external urethral opening. The striated muscle of the external urethral sphincter is controlled by the somatic nervous system, so with the exception of the very young, the very old, or people suffering from injury or illness, it is under conscious, voluntary control and can be held in a contracted state until the person is ready to urinate.

URINALYSIS

Being asked to provide a urine sample is a common part of a medical check-up. The most common urine test is called urinalysis. The sample may be analysed by sight and smell, and with simple urine test strips. The colour and clarity of urine can be the first indicators of abnormalities. Normal urine is yellow to amber in colour and looks clear. If it is very diluted, normal urine may have virtually no odour. Concentrated urine will have a stronger odour. Urine with a sweet smell may be the result of sugar in the urine, a sign of diabetes. Nearly colourless urine could be a sign of excessive fluid intake, or it might be an indicator of diabetes. Very dark urine may signal dehydration but may also be the result of taking certain medications or something in the diet. A reddish tinge is often a sign of blood in the urine, which could result from a urinary tract infection, a kidney stone or even cancer. Cloudy urine might indicate white blood cells in the urine, another sign of a urinary tract infection.

CHAPTER TEN

The Respiratory System

The Respiratory System

You may survive weeks without food, days without water, but you will only last minutes without oxygen. Respiration is the vital process, carried out by the respiratory system, by which the body exchanges gases with the outside world. It is not the same as cellular respiration, the metabolic process that takes place inside cells by which they obtain their energy, but the two processes are linked. Aerobic cellular respiration uses oxygen and generates carbon dioxide as a waste product. It is the job of the respiratory system to supply one and dispose of the other.

Air flows into and out of the body through a continuous system of passages called the respiratory tract. The organs of the upper respiratory tract are the nasal cavity, pharynx and larynx, while the trachea, bronchi and wings make up the lower respiratory tract. Certain muscles of the thorax (chest cavity) also play an important role in breathing, particularly the diaphragm and the muscles between the ribs.

Breathing is a two-step process that involves drawing air into the lungs (inhaling) and letting it back out again (exhaling). Inhaling results

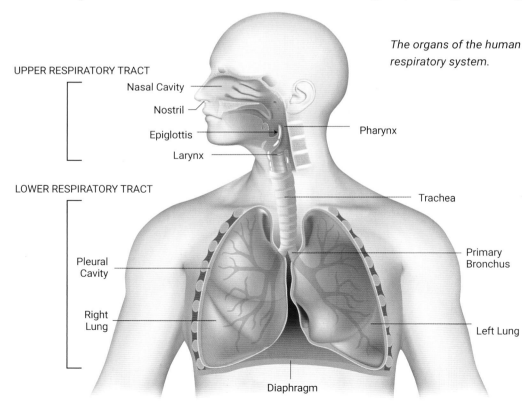

The organs of the human respiratory system.

UPPER RESPIRATORY TRACT

Nasal Cavity

Nostril

Epiglottis

Larynx

Pharynx

LOWER RESPIRATORY TRACT

Trachea

Pleural Cavity

Primary Bronchus

Right Lung

Left Lung

Diaphragm

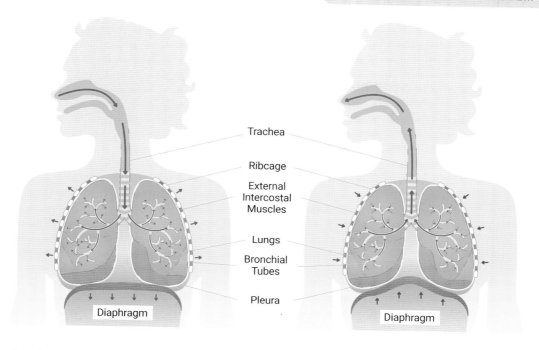

Trachea

Ribcage

External
Intercostal
Muscles

Lungs

Bronchial
Tubes

Pleura

Diaphragm

Diaphragm

Breathing is a two-step process of inspiration, or inhaling, and expiration, or exhaling. During inhalation the downward movement of the diaphragm and expansion of the ribcage by the intercostal muscles causes air to flow in to the lungs. During exhalation, the diaphragm moves up, the ribcage contracts, and air is forced back out again.

mainly from the contraction of the diaphragm, a large, dome-shaped muscle below the lungs that separates the thoracic and abdominal cavities. When the diaphragm contracts it moves down, expanding the thoracic cavity, and pushing the contents of the abdomen downwards. The intercostal muscles between the ribs also contribute to the process of inhalation, especially when taking a deep breath, by expanding the ribs outwards. The resulting increase in thoracic volume causes a decrease in thoracic air pressure. Because the air pressure inside the lungs is lower than that outside the body, outside air flows into the lungs through the respiratory tract.

When the diaphragm relaxes again it moves upwards, reducing the volume of the thorax, with

the result that the air pressure inside the lungs increases, so it is now higher than the external air pressure. With the change in air pressure the lungs contract, forcing out, or exhaling, some of the air they contain.

Breathing may be through the nose or the mouth, although nasal breathing has certain advantages over mouth breathing. The hair-lined nasal passages are better at filtering particles out of the air and are also better at warming and moistening the air, an advantage in winter when the air is cold and dry.

Both inhalation and exhalation can be consciously controlled although most of the time breathing continues without us having to think about it. You can hold your breath for a time if you wish but it is impossible to voluntarily stop breathing indefinitely. It is also possible to inhale and exhale more forcefully than normal, for example while singing or playing a woodwind instrument, or blowing up a balloon. Unconscious control of breathing is mediated by respiratory centres in the medulla and pons of the brainstem, which automatically regulate the rate of breathing based on the body's needs.

THE UPPER RESPIRATORY TRACT

The organs of the upper respiratory tract form a pathway through which air moves into and out of the body. They clean, humidify and warm the incoming air before it reaches the lungs.

The nasal cavity opens to the outside world through the two nostrils. It is a large, air-filled space in the skull located in the middle of the face, above and behind the nose. As inhaled air flows through the nasal cavity, it is warmed and humidified by blood vessels very close to the surface of the epithelial tissue that lines the nasal cavity. Larger foreign particles in the air are trapped by hairs in the nose and by mucus produced by the mucous membranes, preventing them from travelling deeper into the respiratory tract. As well as its function in the respiratory system, the nasal cavity also plays a role in the senses of smell and taste (see pages 129–32).

The tube-like pharynx leads down into the throat from the nasal cavity and the back of the mouth. Both air en route to the lower respiratory tract and food on its way into the digestive system pass through the pharynx. Air passes from the nasal cavity through the pharynx to the larynx while food passes through to the oesophagus. When swallowing occurs, the backward motion of the tongue closes the epiglottis over the entrance to the larynx, preventing swallowed material from moving deeper into the respiratory tract. If any material does make its way into the larynx, it triggers a strong cough reflex that expels it back out.

The larynx is also commonly known as the voice box because it contains the vocal cords, which vibrate when air flows over them, thereby producing sound. Different muscles in the larynx move the vocal cords apart to allow breathing and together to allow the production of vocal sounds as well as adjust their pitch and volume.

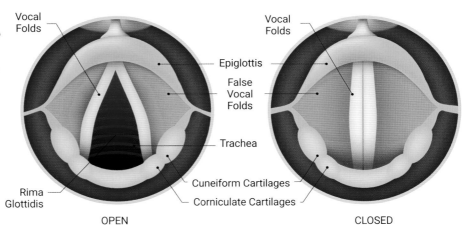

The larynx, or voice box, contains the vocal cords. These vibrate when air flows over them. They can be moved apart to allow breathing, or moved together to produce sounds.

Vocal Folds

Vocal Folds

Epiglottis

False Vocal Folds

Trachea

Cuneiform Cartilages

Corniculate Cartilages

Rima Glottidis

OPEN

CLOSED

THE LOWER RESPIRATORY TRACT

The trachea and other passages of the lower respiratory tract conduct air between the upper respiratory tract and the lungs.

The 10 to 15 cm (4–6 in) long trachea, or windpipe, connects the larynx to the lungs. It is formed by rings of cartilage, giving it strength and resilience. It is the widest passageway in the respiratory tract – about 2.5 cm (1 in) wide – and branches to form two bronchial tubes, the right and left bronchi. These passages branch repeatedly as they move deeper into the lungs, forming a shape reminiscent of an upside-down tree. Each bronchus divides into smaller, secondary bronchi which in turn branch into still smaller tertiary bronchi. The smallest bronchi branch into tiny tubules called bronchioles. So extensive is this branching that there are around 2,400 km (1,500 miles) of airways conducting air through the human respiratory tract. The smallest of the bronchioles end in alveolar ducts, which terminate in clusters of minuscule air sacs, called alveoli, in the lungs.

The lungs, located within the pleural cavity of the thorax, are the largest organs of the respiratory tract. They are surrounded by two thin membranes called pleura, which secrete a fluid that allows the lungs to move freely, expanding

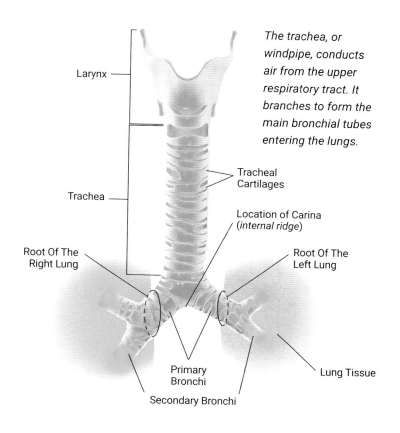

The trachea, or windpipe, conducts air from the upper respiratory tract. It branches to form the main bronchial tubes entering the lungs.

Larynx

Trachea

Tracheal Cartilages

Location of Carina (*internal ridge*)

Root Of The Right Lung

Root Of The Left Lung

Primary Bronchi

Secondary Bronchi

Lung Tissue

and contracting within the pleural cavity. Each of the two lungs is divided into sections called lobes, separated from each other by connective tissues. The larger right lung has three lobes, while the smaller left lung contains only two lobes. The left lung is smaller to accommodate the heart, which is just left of the centre in the chest.

The lungs receive deoxygenated blood from the right side of the heart by way of the pulmonary artery. This blood absorbs oxygen in the lungs and carries it back to the left side of the heart through pulmonary veins. There are four pulmonary veins (two for each lung), and all four carry oxygenated blood to the heart. The oxygenated blood is then pumped to cells

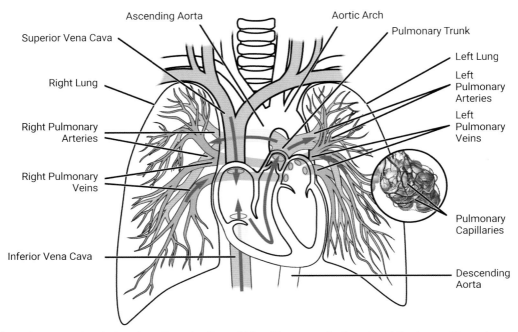

The pulmonary circulation describes the flow of blood between the heart and the lungs. The pulmonary artery carries blood to the lungs, where it gives up carbon dioxide and picks up oxygen. The oxygenated blood travels back through the four pulmonary veins to the left atrium of the heart. From the left atrium, the blood is pumped into the left ventricle and through the aortic valve into the aorta, and hence to the rest of the body, via the systemic circulation. The blood returns to the right atrium via the superior vena cava and inferior vena cava, and the cycle repeats.

throughout the body, including to the cells of the lungs, for cellular respiration.

Alveoli

Lung tissue consists mainly of alveoli, tiny sacs of connective tissue and simple squamous epithelial tissue. The connective tissue in the alveoli includes elastic fibres that allow them to stretch and expand as they fill with air during inhalation. During exhalation, the fibres spring back and expel the air. Cells in the walls of the alveoli secrete a film of fatty substances called surfactant which prevents the alveolar walls from collapsing and sticking

together when air is expelled. Other cells in alveoli include macrophages, part of the body's immune system which engulf and destroy bacteria and other particles that manage to get into the lungs.

Alveoli are the functional units of the lung, just as nephrons are the functional units of the kidney (see pages 163–4). This is where the vital process of gas exchange takes place with the capillary network that surrounds the alveoli. Between them, the lungs may contain as many as 700 million alveoli, providing a huge total surface area, perhaps as much as 100 m² (1,000 ft²), for gas exchange to take place. The alveoli are arranged in groups that look a little like clusters of grapes. Each alveolus is covered with epithelium that is just one cell thick and is surrounded by a bed of pulmonary capillaries, each of which also has an epithelium one cell thick. This means that gases crossing between an alveolus and its surrounding capillaries only have to traverse two cells.

Gas molecules naturally move down a concentration gradient from an area of higher concentration to an area of lower concentration, a process called diffusion that requires no input of energy to take place. When you inhale there is a greater concentration of oxygen in the air in the alveoli than there is in the blood in the pulmonary capillaries. As a result, oxygen diffuses from the air inside the alveoli into the blood in the surrounding mesh of capillaries. At the same time the concentration of carbon dioxide is greater in the blood in the pulmonary capillaries than it is inside the alveoli with the result that carbon dioxide diffuses in the opposite direction.

The oxygenated blood reaching the cells of the body has a much higher concentration of oxygen than there is within the cells and as a result, oxygen diffuses from the capillaries into the cells. By contrast there is a much higher concentration of carbon dioxide in the cells which diffuses from the cells into the peripheral capillaries to be carried to the lungs for elimination.

The respiratory system works with the nervous and cardiovascular systems to maintain homeostasis. Levels of oxygen and carbon dioxide in the blood must be maintained within a limited range to ensure survival. Cells cannot carry out aerobic respiration without oxygen, and too much carbon dioxide in the blood results in it becoming dangerously acidic, while too little carbon dioxide makes it too alkaline. The level of carbon dioxide in the blood is detected by the brain, which instructs the autonomic nervous system to speed up or slow down the rate of breathing as needed to bring the carbon dioxide level within the optimum range. Faster breathing lowers the carbon dioxide level, at the same time raising the oxygen level and blood pH, while slower breathing has the opposite effect.

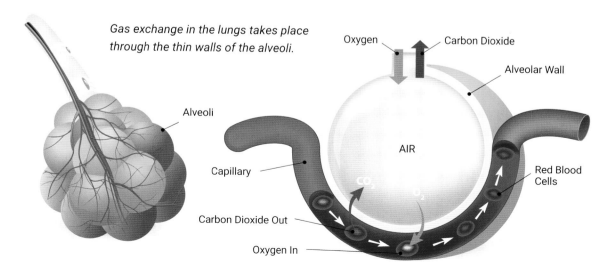

Gas exchange in the lungs takes place through the thin walls of the alveoli.

Oxygen

Carbon Dioxide

Alveolar Wall

Alveoli

AIR

Capillary

Red Blood Cells

Carbon Dioxide Out

Oxygen In

The Cardiovascular System

CHAPTER ELEVEN

The Cardiovascular System

The cardiovascular, or circulatory, system, transports materials, including oxygen from the lungs, nutrients from the digestive system and hormones from the endocrine system to all the cells of the body. It also takes waste materials from the cells to the respiratory and renal systems for elimination. The main components of the cardiovascular system are the heart, blood vessels and the blood that flows through them.

The blood vessels of the cardiovascular system form a complex interconnected network through which blood flows via two linked one-way circuits.

The heart, the lungs and the major blood vessels that connect them form the pulmonary circuit. The right ventricle of the heart pumps deoxygenated blood into the right and left pulmonary arteries which carry the blood to the right and left lungs, respectively. The deoxygenated blood picks up oxygen from the alveoli in the lungs, at the same time giving up carbon dioxide. As a result, the blood returning to the heart is almost completely saturated with oxygen. The oxygenated blood returns from the right and left lungs through the two right and two left pulmonary veins. All four pulmonary veins enter the left atrium of the heart.

The oxygenated blood in the left atrium of the heart moves to the left ventricle which pumps it directly into the largest artery in the body, the aorta. The blood is now in the systemic circuit, the part of the cardiovascular system that transports blood to and from all of the tissues of the body to provide oxygen and nutrients, and to pick up wastes. The systemic circuit is formed of the heart and blood vessels that supply the metabolic

COMPONENTS OF BLOOD

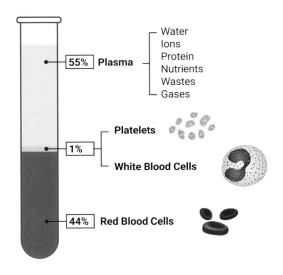

needs of all the cells in the body, including those of the heart and lungs.

Major arteries branching off the aorta carry the blood to the head and upper extremities and to the abdomen and lower extremities. The blood then returns to the heart through a network of increasingly larger veins eventually collecting in the superior vena cava (from the upper body) and inferior vena cava (from the lower body), which empty directly into the right atrium of the heart.

Opposite: The major arteries (in red) and veins (in blue) of the human circulatory system.

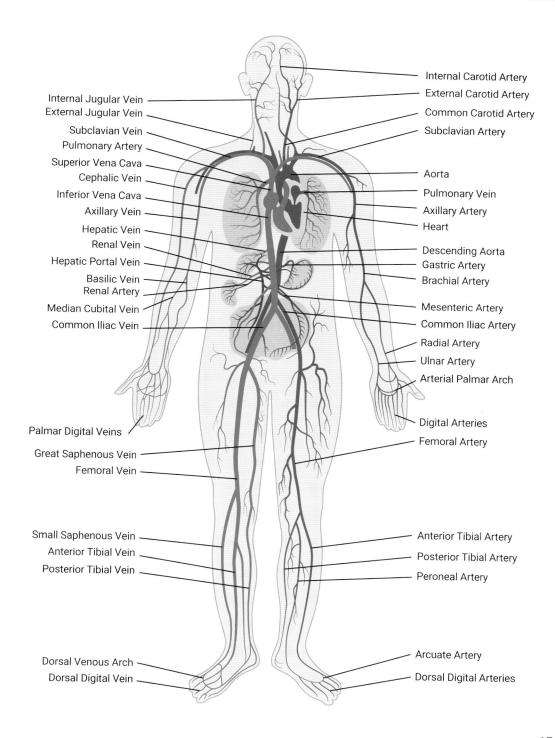

Internal Carotid Artery
External Carotid Artery
Common Carotid Artery
Subclavian Artery

Internal Jugular Vein
External Jugular Vein
Subclavian Vein
Pulmonary Artery
Superior Vena Cava
Cephalic Vein
Inferior Vena Cava
Axillary Vein
Hepatic Vein
Renal Vein
Hepatic Portal Vein
Basilic Vein
Renal Artery
Median Cubital Vein
Common Iliac Vein

Aorta
Pulmonary Vein
Axillary Artery
Heart
Descending Aorta
Gastric Artery
Brachial Artery
Mesenteric Artery
Common Iliac Artery
Radial Artery
Ulnar Artery
Arterial Palmar Arch
Digital Arteries
Femoral Artery

Palmar Digital Veins
Great Saphenous Vein
Femoral Vein

Small Saphenous Vein
Anterior Tibial Vein
Posterior Tibial Vein

Anterior Tibial Artery
Posterior Tibial Artery
Peroneal Artery

Dorsal Venous Arch
Dorsal Digital Vein

Arcuate Artery
Dorsal Digital Arteries

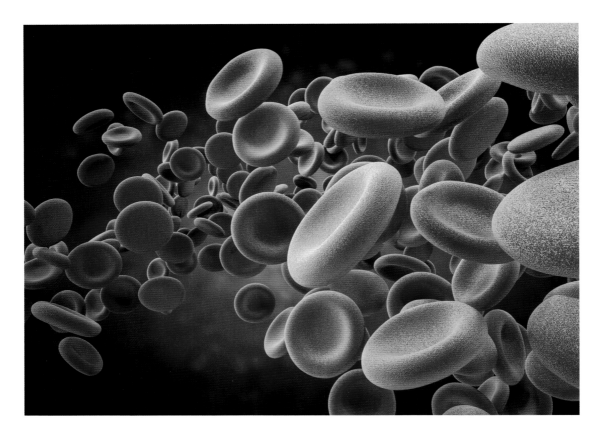

Red blood cells, or erythrocytes. Their red colour comes from haemoglobin, an iron containing protein that binds with oxygen.

BLOOD

The function of the cardiovascular system is to transport blood around the body. So, what is blood? Blood is a fluid connective tissue, a connective tissue in which the cells aren't actually connected to each other. The average adult body contains between 4.7 and 5.7 litres (160–193 fluid oz) of blood. More than half of that amount is fluid; most of the rest consists of blood cells.

The liquid component of blood is called plasma. It is about 92 per cent water and makes up about 55 per cent of blood by volume. Plasma contains many dissolved substances, mostly proteins, but with trace amounts of glucose, mineral ions, hormones, carbon dioxide and other substances.

Plasma also contains blood cells. These are erythrocytes, leukocytes and thrombocytes. All three types of blood cells arise from stem cells in red marrow within bones in a process called haematopoiesis.

The most numerous of the blood cells are the erythrocytes, more commonly known as red blood cells. By number, red blood cells make up more than three-quarters of all the cells in the human body. One microlitre of blood contains between four and six million red blood cells. The cytoplasm of a mature erythrocyte is almost completely filled with haemoglobin, the iron-containing protein that binds with oxygen and gives the cell its red colour. To maximize the space available for haemoglobin, mature erythrocytes lack a cell nucleus.

Leukocytes, or white blood cells, are part of the body's immune system (see page 194), attacking invading pathogens and removing old cells. There are different types of leukocyte, each with a different immune function. There are far fewer leukocytes than there are red blood cells, normally only about 4,000 to 10,000 per microlitre of blood. Unlike erythrocytes, leukocytes have a nucleus.

Thrombocytes, or platelets, are the smallest of the blood cells. They are more numerous than white blood cells and, like erythrocytes, they have no nucleus. There are about 150,000 to 400,000 thrombocytes per microlitre of blood. The main function of thrombocytes is in haemostasis, the control of bleeding. When a blood vessel is damaged large numbers of platelets attach to the injured surface and to each other, forming a sticky mass which stops the bleeding and begins to form a blood clot, or thrombus. The thrombocytes also release substances into the blood that initiate the formation of the protein fibrin. Strands of fibrin crisscross and strengthen the platelet plug.

Haemostasis is the body process that controls bleeding, involving the restriction of blood flow and the formation of a clot in the injured blood vessel.

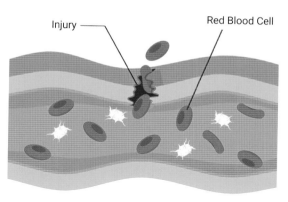

VASOCONSTRICTION

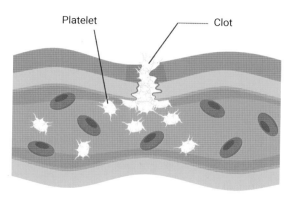

PRIMARY HAEMOSTASIS

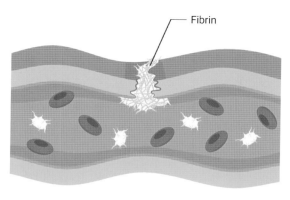

CLOT FORMATION

BLOOD TYPE

Erythrocytes carry antigens, which are like markers on the outside of the cell, that determine blood type. These types, or groups, differ between individuals – the most common groups being designated A, B and O. Knowing what group a person belongs to is very important if that person needs a blood transfusion. If, for example, blood from someone who belongs to the A or B group is introduced into the body of someone whose blood is of the opposite group, it will trigger an immune response. The recipient's immune system will attack the alien blood cells, causing them to burst.

Blood groups, or types, are a genetic characteristic determined by the presence of molecules called antigens on the surface of red blood cells. If you are given a blood transfusion from someone whose blood type is not the same as your own, the immune system produces antibodies that mark out the alien blood cells for destruction.

	Group A	Group B	Group AB	Group O
Red Blood Cell Type	A	B	AB	O
Antibodies in Plasma	Anti-B	Anti-A	None	Anti-A and Anti-B
Antigens In Red Blood Cell	A Antigen	B Antigen	A and B Antigens	None

THE HEART

The function of the heart is to pump blood through the vessels of the cardiovascular system. It is about the size of a fist in a healthy adult, shaped something like an upside-down pear, and is located behind the sternum, slightly to the left of centre in the chest. The average adult female heart weighs around 250 to 300 grams (9–10.5 oz), that of an average adult male around 300 to 350 grams (10.5–12 oz). The heart of an athlete can be much larger than this.

The wall of the heart is thick and muscular. Internally, it is divided into four chambers through which blood flows. The wall of the heart has three layers: the endocardium, the myocardium and the pericardium. The endocardium is the

The anatomy of the heart.

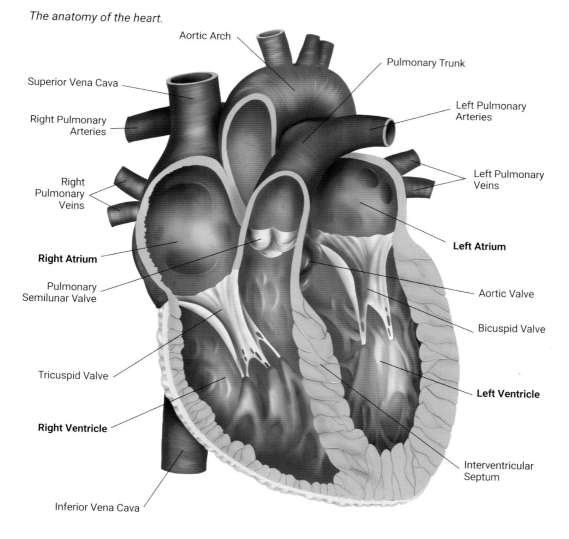

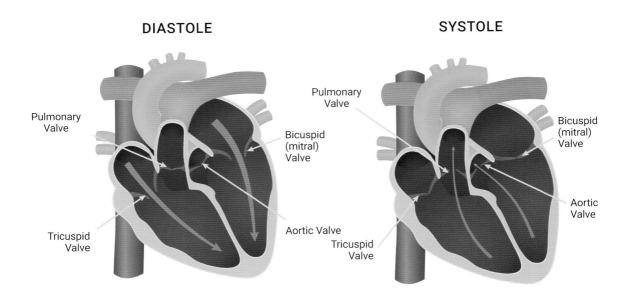

DIASTOLE

SYSTOLE

Pulmonary
Valve

Bicuspid
(mitral)
Valve

Tricuspid
Valve

Aortic Valve

Pulmonary
Valve

Bicuspid
(mitral)
Valve

Aortic
Valve

Tricuspid
Valve

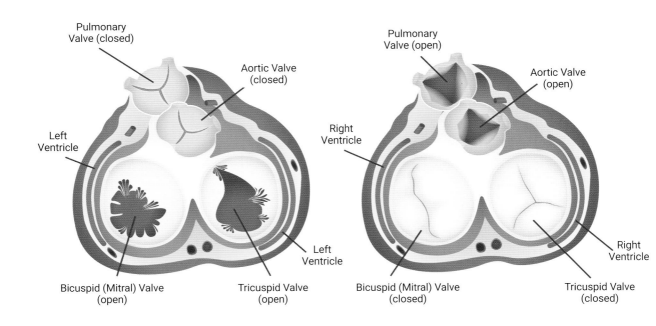

Pulmonary
Valve (closed)

Aortic Valve
(closed)

Left
Ventricle

Left
Ventricle

Bicuspid (Mitral) Valve
(open)

Tricuspid Valve
(open)

Pulmonary
Valve (open)

Aortic Valve
(open)

Right
Ventricle

Right
Ventricle

Bicuspid (Mitral) Valve
(closed)

Tricuspid Valve
(closed)

innermost layer of the heart wall. It consists primarily of simple epithelial cells and covers the heart chambers and valves. A thin layer of connective tissue joins the endocardium to the myocardium, the middle and thickest layer of the heart wall. The myocardium is cardiac muscle surrounded by a framework of collagen.

There are two types of cardiac muscle cells in the myocardium: cardiomyocytes, which make up about 99 per cent of the cardiac muscle, and pacemaker cells, which conduct electrical impulses that cause the cardiomyocytes to contract. The myocardium is supplied with blood vessels and nerve fibres via the pericardium, a protective sac that encloses and protects the heart. The pericardium consists of the visceral pericardium and parietal pericardium membranes. A fluid-filled cavity between the two membranes helps both to cushion the heart and to lubricate its outer surface.

The four internal chambers of the heart are separated from each other by dense connective tissue consisting mainly of collagen. They include two upper atria (singular: atrium), and two lower ventricles. Blood coming into the heart first enters the atria. The right atrium receives deoxygenated blood through the superior vena cava (from the upper body) and inferior vena cava (from the lower body). The left atrium receives oxygenated blood from the lungs through the pulmonary veins. The right ventricle discharges blood to the lungs through the pulmonary artery, and the left ventricle discharges blood to the rest of the body through the aorta.

Heart valves ensure that blood can only flow through the heart in one direction: from the atria to the ventricles, and from the ventricles to the pulmonary artery and aorta. The tricuspid atrioventricular valve allows blood to flow from the right atrium to the right ventricle. The bicuspid atrioventricular valve (also known as the mitral valve) leads from the left atrium to the left ventricle. The pulmonary semilunar valve controls flow from the right ventricle to the pulmonary artery and the aortic semilunar valve permits flow from the left ventricle to the aorta.

The atrioventricular valves prevent back flow when the ventricles are contracting, while the semilunar valves prevent back flow from the pulmonary artery and aorta. The atrioventricular valves have to withstand considerable force when the ventricles contract. To do so they are reinforced with structures called chordae tendineae. These tendon-like cords of connective tissue anchor the valve and prevent it from opening in the wrong direction. The chordae tendineae are themselves anchored to the interior of the ventricles by the papillary muscles, which are specialized for the task.

The deoxygenated blood that collects in the right atrium is pumped through the tricuspid valve into the right ventricle. From the right ventricle, the blood is pumped through the pulmonary valve into the pulmonary artery which carries the blood to the lungs, where it gives up carbon dioxide and picks up oxygen. Oxygenated blood travels back from the lungs through the pulmonary veins, entering the left

Opposite: The valves of the heart ensure that blood can only flow through in one direction, opening to allow blood into the appropriate chamber and then closing again to prevent any back flow.

THE CARDIAC CYCLE

During the cardiac cycle, the atria and ventricles work together so that blood is pumped efficiently through and out of the heart. There are two parts to the cycle: diastole and systole. During diastole, the atria contract and pump blood into the ventricles, while the ventricles relax as they fill with blood. During systole, the atria relax and take in blood from the lungs and body, while at the same time the ventricles contract and pump blood out of the heart. It is those twin contractions, one after another, that produce the familiar 'lub dub' sound of a healthy, beating heart.

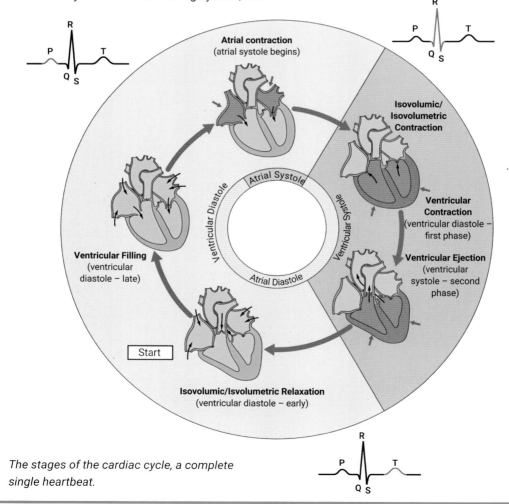

The stages of the cardiac cycle, a complete single heartbeat.

atrium of the heart. From the left atrium, the blood is pumped through the mitral valve into the left ventricle and from there the blood is pumped through the aortic valve into the aorta, which branches into smaller arteries carrying oxygenated blood to the rest of the body. After passing through capillaries and exchanging substances with cells, the blood returns to the right atrium via the superior vena cava and inferior vena cava, and another cycle begins.

Sinus rhythm

The normal, rhythmical beating of the heart is called sinus rhythm. It is maintained by pacemaker cells, which are located in an area of the heart called the sinoatrial node. For each cardiac cycle, an electrical signal travels from the pacemaker cells first to the right and left atria stimulating them to contract together. The signal then travels to another node, called the atrioventricular node, and from there to the right

Trace on an electrocardiogram showing the pattern of a normal sinus rhythm, the rhythmical beating of the heart.

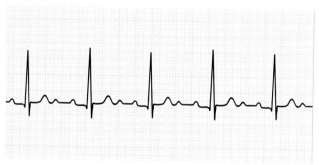

Normal Sinus Rhythm

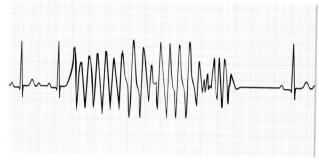

Ventricular Tachycardia to Sinus Rhythm Background

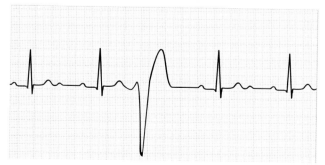

Single Early Ventricular Complex In The Background Of Sinus Rhythm

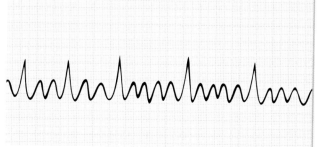

Atrial Fibrillation

and left ventricles, which also contract together a fraction of a second after the atria contract.

Two paired cardiovascular centres in the medulla of the brainstem influence the normal sinus rhythm of the heart through sympathetic and parasympathetic nerves as part of the autonomic nervous system. The parasympathetic nerves act to decrease the heart rate, and the sympathetic nerves act to increase it. Without the parasympathetic input, the pacemaker cells would generate a resting heart rate of around 100 beats per minute, instead of the more normal resting heart rate of around 72 beats per minute.

The cardiovascular centres receive input from receptors throughout the body, and adjust the heart rate, as needed. Increased physical activity, for example, triggers receptors in muscles, joints and tendons, which send nerve impulses to the cardiovascular centres, which increase the heart rate. Other factors can also affect the

The structure of the body's blood vessels.

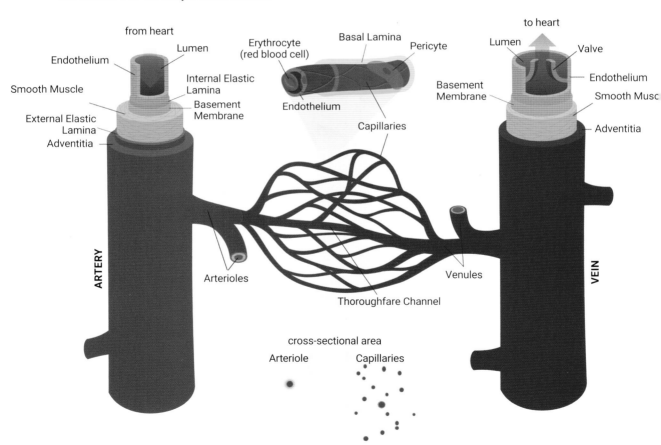

heart rate. Feeling relaxed and at rest contributes to a decrease in the heart rate. Dehydration or overheating can cause the heart to beat faster, as can the release of hormones such as the adrenal hormones when we are frightened.

In common with all other organs in the body, the heart itself requires a blood supply. The cardiomyocytes are continually active, responsible for the constant beating of the heart, and need a continuous supply of oxygen and nutrients, as well as having the carbon dioxide and waste products they produce removed. The blood vessels that carry blood to and from the heart muscle cells make up the coronary circulation. The coronary circulation supplying the heart tissues is distinct from the general circulation carrying blood from the heart to the other organs of the body and back again. The coronary arteries supply oxygen-rich blood to the heart muscle cells, the right coronary artery supplies the right side of the heart, and the left coronary artery the left side of the heart. The coronary veins remove deoxygenated blood and wastes, draining into the right atrium.

BLOOD VESSELS

Blood is transported around the body through a complex network of blood vessels. There are three major types: arteries, veins and capillaries.

An artery is defined as a blood vessel that carries blood away from the heart. An exception to this rule is the coronary arteries which supply blood to the heart muscle cells as they travel towards the heart, not away from it. The pumping action of the heart provides the pressure that makes the blood flow through the arteries. Most arteries, including the coronary arteries, carry oxygenated blood to the cells of the body. The pulmonary artery is different in that it carries

deoxygenated blood from the heart to the lungs, where it picks up oxygen and releases carbon dioxide. In most cases, arterial blood is highly saturated with oxygen (95–100 per cent) stored in the haemoglobin in red blood cells.

The largest artery in the body is the aorta which has high-pressure, oxygenated blood pumped directly into it from the left ventricle of the heart. The aorta extends down into the abdomen and has many branches, which subdivide repeatedly, growing smaller and smaller in diameter. The smallest arteries are called arterioles.

A vein is a blood vessel that carries blood towards the heart. With the exception of the pulmonary veins carrying oxygenated blood from the lungs to the heart, most veins carry deoxygenated blood back to the heart. Blood is pushed through the veins by movements of the skeletal muscles – it is not under pressure in the same way as arterial blood, pushed by the heart. Venous blood is prevented from flowing backwards by valves in the larger veins. At any one time about 60 per cent of the body's blood is contained within veins.

The largest veins in the body are the superior vena cava, carrying blood from the upper body to the right atrium of the heart, and the inferior vena cava, carrying blood from the lower body, also to the right atrium. Like arteries, veins form a complex, branching system of larger and smaller vessels. The smallest veins are called venules.

Arterioles and venules are linked together by the network of capillaries, the smallest blood vessels in the cardiovascular system. Only a single red blood cell at a time can squeeze through a capillary. Capillaries generally form a branching network of vessels, called a capillary bed, that provides a large surface area for the exchange

BLOOD PRESSURE

Blood pressure is a measure of the force that blood exerts on the walls of arteries. It is highest when the heart contracts and pumps out blood, and lowest when the heart relaxes and refills with blood. It is generally measured in millimetres of mercury (mm Hg), and expressed as a double number, a higher number for systolic pressure when the ventricles contract, and a lower number for diastolic pressure when the ventricles relax. Normal blood pressure is generally defined as less than 120 mm Hg (systolic)/80 mm Hg (diastolic) when measured in the arm at the level of the heart. It decreases as blood flows further away from the heart and into the smaller arteries as increasing friction of the blood against the arterial walls leads to greater resistance to the blood flow. This resistance reduces blood pressure before the blood reaches the thin-walled capillaries which cannot withstand great pressure. The pressure of blood against the walls of veins is greatly reduced, normally no more than 10 mm Hg.

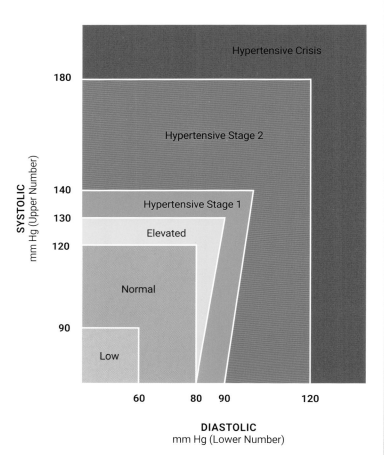

Blood pressure measures the force exerted by blood as it flows through the arteries; systolic pressure is recorded when the ventricles contract pumping blood out of the heart, and diastolic is recorded when they relax. High blood pressure, or hypertension, is a health risk.

VASOCONSTRICTION AND VASODILATION

Smooth muscles in the walls of arteries can contract to narrow the lumen (vasoconstriction) or relax to allow it to widen (vasodilation). This helps to regulate blood pressure. Narrowing of the blood vessels increases friction with the arterial walls, reducing flow and resulting in a drop in blood pressure. The smooth muscles are controlled by the autonomic nervous system in response to pressure-sensitive sensory receptors in the walls of larger arteries. Vasoconstriction and vasodilation also play an important part in the fight-or-flight response. Vasodilation increases the blood flow to skeletal muscles, while the flow to the digestive organs is reduced by vasoconstriction. These responses are under the control of the sympathetic nervous system and triggered by the release of hormones. Arteries can also dilate or constrict to help regulate body temperature by allowing more or less blood to flow to the body's surface.

of substances between the blood and the cells of the surrounding tissues.

All blood vessels are basically hollow tubes with an internal space, called a lumen, through which the blood flows. The walls of capillaries are extremely thin, consisting of a single layer of epithelial cells to permit the exchange of substances, including water, oxygen and nutrients, as well as waste products, between the blood within the capillaries and the cells of surrounding tissues.

Both arteries and veins have walls comprising three layers: the tunica intima, tunica media and tunica externa (or tunica adventitia). The tunica intima, the innermost and also the thinnest layer, consists of a single layer of endothelial cells surrounded by a thin layer of connective tissues. It reduces friction between the blood and walls of the blood vessel. The middle layer, the tunica media, is the thickest layer in arteries and consists mainly of elastic fibres and connective tissues. In arteries, it also contains smooth muscle tissues, which control the diameter of the vessels. The outer layer, the tunica externa, consists of connective tissue and nerves and provides strength and protection. This is the thickest layer in veins.

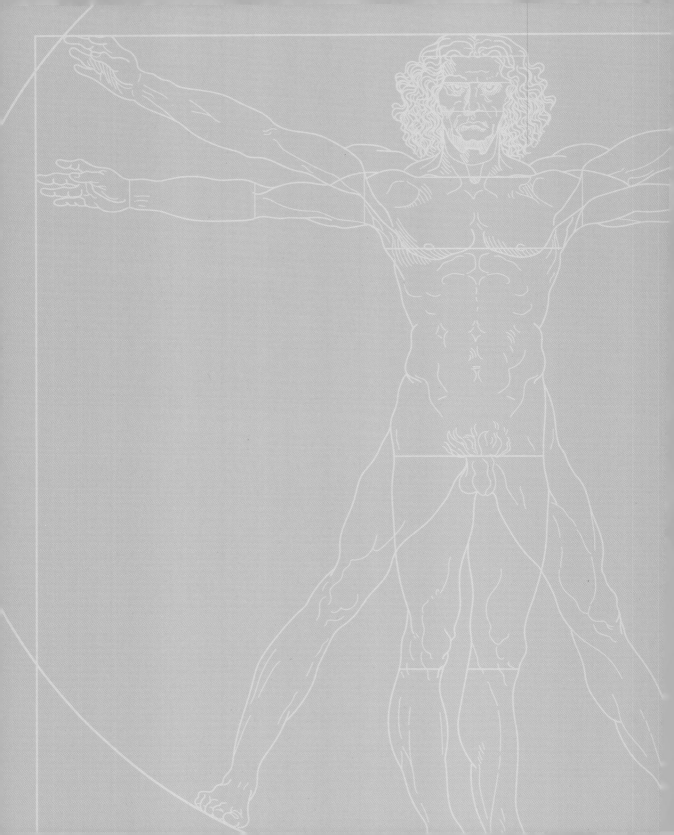

The Immune System

CHAPTER TWELVE

The Immune System

The immune system is a multi-layered defence network protecting the body from attack by disease-causing pathogens, including viruses, bacteria, fungi and parasites. It is activated when the body detects the presence of something that is not a part of itself.

The immune system responds when it is triggered by antigens. Antigens are molecules that the immune system identifies as either self (that is, produced by your own body) or non-self (not produced by your own body). Antigens may be proteins on the surface of a virus, carbohydrates or other chemicals. If antigens are identified as non-self, the immune system reacts by forming antibodies that are specific to the non-self antigens.

THE LYMPHATIC SYSTEM

The lymphatic system, which also plays a part in the digestive and circulatory systems, is a major

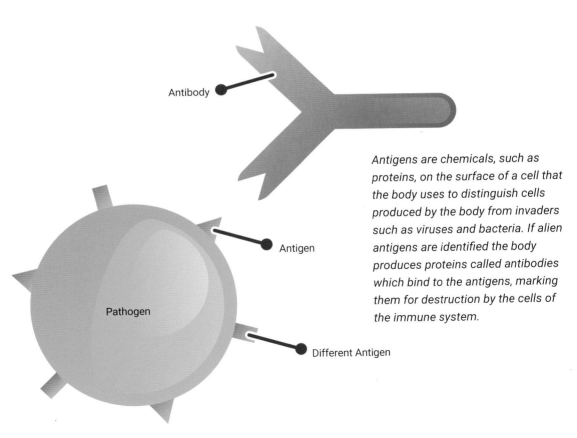

Antibody

Antigen

Pathogen

Different Antigen

Antigens are chemicals, such as proteins, on the surface of a cell that the body uses to distinguish cells produced by the body from invaders such as viruses and bacteria. If alien antigens are identified the body produces proteins called antibodies which bind to the antigens, marking them for destruction by the cells of the immune system.

component of the immune system. Organs of the lymphatic system include the tonsils, the thymus, the spleen and hundreds of lymph nodes distributed along the lymphatic vessels.

When blood travels through the cardiovascular system some water and nutrients pass through the walls of the capillaries and into the tissue spaces between cells, forming interstitial fluid which nourishes the cells and also absorbs their waste products. Much of the water in the

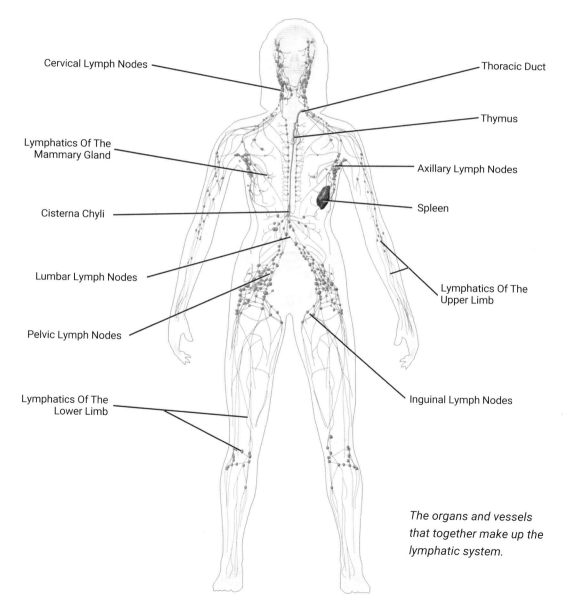

Cervical Lymph Nodes

Thoracic Duct

Thymus

Lymphatics Of The Mammary Gland

Axillary Lymph Nodes

Cisterna Chyli

Spleen

Lumbar Lymph Nodes

Lymphatics Of The Upper Limb

Pelvic Lymph Nodes

Lymphatics Of The Lower Limb

Inguinal Lymph Nodes

The organs and vessels that together make up the lymphatic system.

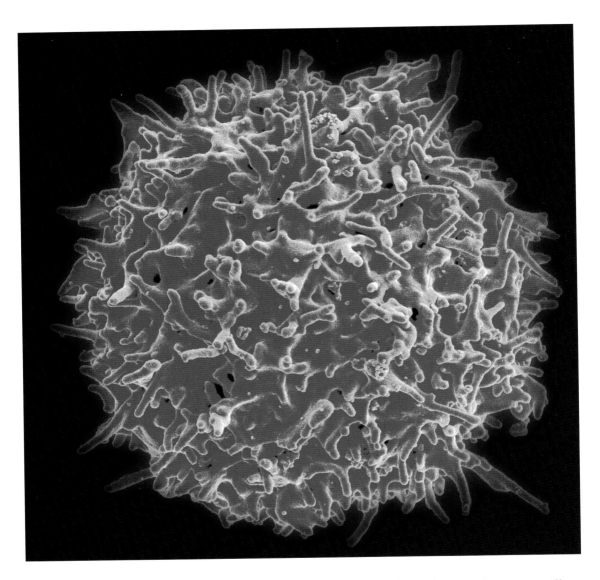

An electronmicrograph of a human T cell, a type of lymphocyte that matures in the thymus. It is one of the main lines of defence in the immune system, attacking particular pathogens that enter the body.

interstitial fluid is reabsorbed into the capillary blood by osmosis. Most of the remaining fluid is absorbed by tiny lymphatic vessels called lymph capillaries, becoming lymph. Lymph is very similar in composition to blood plasma and may contain proteins, cell waste products, cellular debris and pathogens. It also contains numerous white blood cells, especially lymphocytes which are the main cellular components of lymph.

The lymph is transported through the lymphatic vessel network to two large lymphatic ducts in the upper chest. From there, it flows into two major veins, the subclavian veins of the cardiovascular system. In a similar way to the movement of blood through the veins, lymph moves through the lymphatic vessels via a combination of contractions of the vessels themselves and squeezing of the vessels externally by skeletal muscles. Also, like veins, lymphatic vessels have numerous valves that ensure the lymph flows in just one direction.

The tiny villi in the lining of the small intestine each have an internal network of capillaries and lymphatic vessels called lacteals. The lacteals absorb mainly fatty acids from the digestion of lipids, forming a fluid called chyle. The lymphatic network transports chyle from the small intestine to the blood circulation via the lymphatic ducts and from there it is taken to the liver for processing.

The lymphatic system's primary role within the immune system is to protect the body against pathogens and cancerous cells. This function is carried out by lymphocytes, a type of white blood cell involved in the adaptive immune system, which recognize and defend against specific pathogens or cancerous cells. There are two major types of lymphocytes, called B cells and T cells.

In common with other blood cells, both B cells and T cells are produced from stem cells in the red marrow inside bones. After lymphocytes first form, they must go through a process of maturation during which they 'learn' to distinguish invaders from the components of the body, that is to say self from non-self. B cells are so called because they mature in the bone marrow.

After maturing, they first travel to the circulatory system before entering the lymphatic system. T cells are so named because they mature in the thymus, a small lymphatic organ in the chest.

Both bone marrow and the thymus, where lymphocytes are produced, are called primary lymphoid organs. The tonsils, spleen and lymph nodes, which do not produce lymphocytes, are designated as secondary lymphoid organs. They filter lymph and store lymphocytes. It is in the secondary lymphoid organs that an immune response is triggered by pathogens in which pathogen-specific lymphocytes are cloned and circulated between the lymphatic system and the cardiovascular system, where they search out and destroy their specific pathogens by producing antibodies against them.

WHAT ARE ANTIBODIES?

Antibodies are proteins made by part of the body's immune system called B cells with each B cell producing unique antibodies that recognize a specific antigen. An antigen is any foreign or toxic substance that provokes an immune response. Generally, antibodies are Y-shaped molecules with two sites to which antigens can bind. The normal function of an antibody is to bind antigens and mark them out for destruction by leukocytes. Antibodies are abundant in the mucous membranes lining the respiratory system, the digestive system and the reproductive system, places where the body may come into contact with pathogens in air, food and sexual fluids.

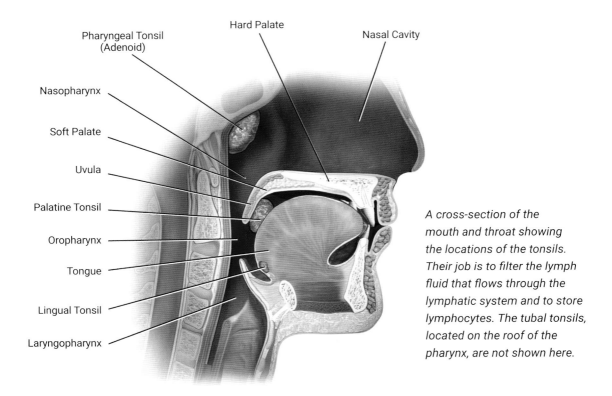

Pharyngeal Tonsil (Adenoid)

Hard Palate

Nasal Cavity

Nasopharynx

Soft Palate

Uvula

Palatine Tonsil

Oropharynx

Tongue

Lingual Tonsil

Laryngopharynx

A cross-section of the mouth and throat showing the locations of the tonsils. Their job is to filter the lymph fluid that flows through the lymphatic system and to store lymphocytes. The tubal tonsils, located on the roof of the pharynx, are not shown here.

Humans have four pairs of tonsils. All four pairs of tonsils encircle the part of the throat where the respiratory and gastrointestinal tracts intersect, a likely pathway for pathogens to enter the body.

The spleen, largest of the secondary lymphoid organs, is centrally located in the body. In addition to storing lymphocytes and filtering lymph, the spleen also filters blood, removing dead or damaged red blood cells. The spleen also produces red blood cells in the foetus, a function that is taken over by bone marrow after birth.

The human body has hundreds of lymph nodes located along the lymphatic vessels, many of them found around the base of the limbs and in the neck. Lymph passes through the nodes on its way back to the blood. Each lymph node contains many lymphocytes and acts like a filter for foreign substances such as infectious bacteria.

THE INNATE IMMUNE SYSTEM

The innate immune system is the body's first line of defence, putting up a number of barriers against invading pathogens. These include physical barriers, such as the skin; chemical barriers, such as the acidic environment of the stomach; and biological barriers, such as white blood cells that ingest and destroy microbes.

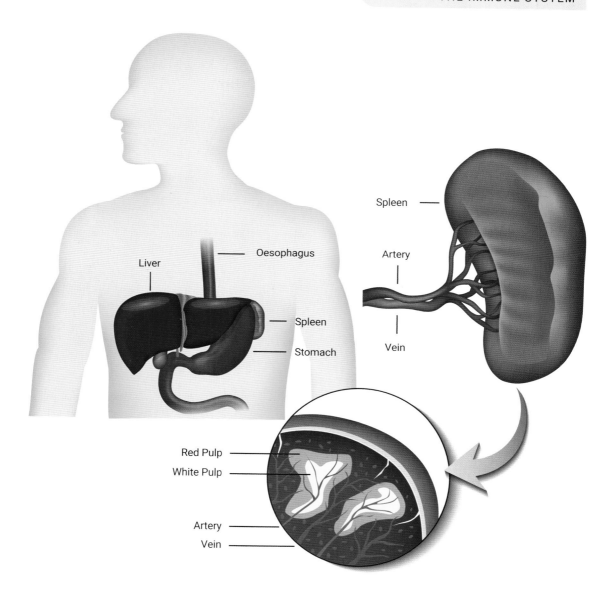

Spleen

Oesophagus

Liver

Spleen

Stomach

Spleen

Artery

Vein

Red Pulp

White Pulp

Artery

Vein

The innate immune system is non-specific in its response to pathogens – it reacts in the same way whatever the pathogen it encounters. If tissue is damaged or infected, leukocytes and defensive proteins flood into the affected areas. These leukocytes and proteins can defend the body from pathogens because they recognize

The anatomy of the spleen and its location in the body. In addition to being one of the secondary organs of the lymphatic system, the spleen also removes dead or damaged red blood cells.

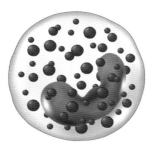  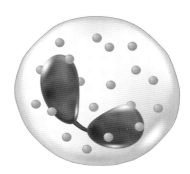

common patterns, or signatures, of molecules that occur in many different types of pathogens. These signature molecules are known as PAMPs, for pathogen-associated molecular patterns. Because PAMPs are commonly found in unrelated pathogens, the leukocytes cannot tell one type of pathogen from another. This non-specific immune response is what is described as innate, or natural, immunity.

The body has significant physical barriers to keep out potential pathogens. Body surfaces, particularly those associated with body openings, are protected by sticky mucus, produced by mucous membranes, which trap pathogens, preventing them reaching deeper into the body. Keratin in the skin also forms an efficient barrier. The skin and mucous membranes also create a hostile chemical environment. The surface of the skin is acidic, which inhibits the growth of bacteria. Mucus, saliva and tears all contain lysozyme, an enzyme that breaks down bacterial cell walls. The highly acidic secretions of the stomach kill many pathogens entering the digestive system.

Biological barriers to infection include the community of microorganisms that make their home on the skin and in the digestive tract. Harmless to us, there is evidence that these organisms are in fact highly beneficial, combating and outcompeting potentially disease-causing organisms.

Pathogens may evade the body's external defences and enter through skin abrasions or other wounds or by being present in numbers too large for the mucous membranes to cope with. If this happens, the innate immune system has a number of internal defences. White blood cells in the blood and lymph recognize pathogens as foreign to the body and respond to them. A white blood cell, or leukocyte, is able to move independently and can leave the blood to go to infected tissues. White blood cells called monocytes circulate in the blood and lymph and develop into macrophages after they move into infected tissue. A macrophage is a large cell capable of engulfing foreign particles and pathogens. The process of ingesting and digesting pathogens is called phagocytosis, and the cells that carry this process out are called phagocytes. When a phagocyte encounters a pathogen, such as a bacterium, it extends a portion of its cell membrane, engulfing the pathogen, which becomes enclosed within an intracellular vesicle where digestive enzymes and acids kill and digest it. The remains of the destroyed pathogen are then ejected by the phagocyte.

Mast cells are a type of leukocyte that take up residence in the tissues other than blood. They

Left: Some of the different types of leukocyte that take part in the innate immune system's response to pathogens.

Below: This electron-micrograph shows bacteria (shown in blue) being engulfed and destroyed by a white blood cell.

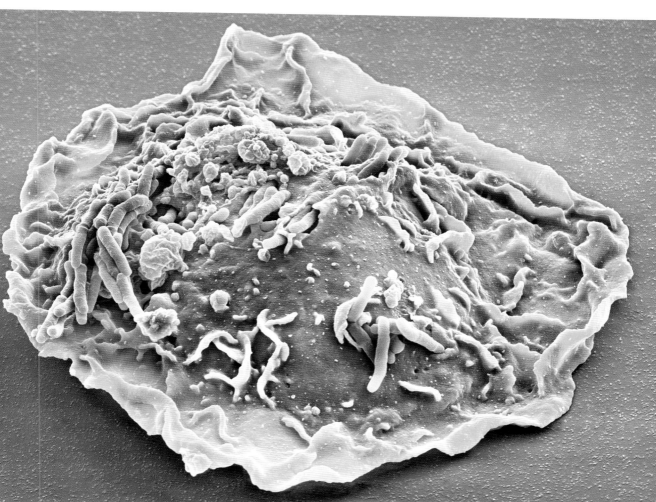

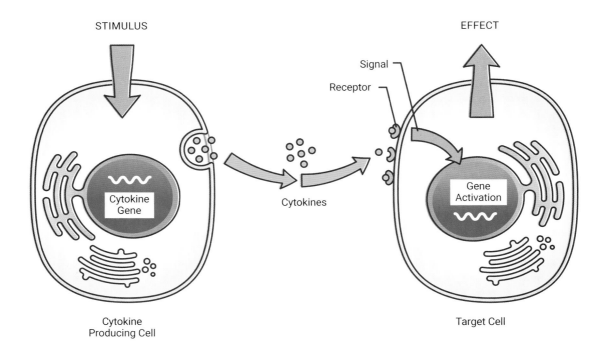

STIMULUS

EFFECT

Signal

Receptor

Cytokine
Gene

Gene
Activation

Cytokines

Cytokine
Producing Cell

Target Cell

*Cytokines are chemicals released by injured or infected cells. They have a
number of effects, including shutting down the processes in infected cells
that viruses need to survive, attracting leukocytes to the site of infection, and
encouraging other cells to produce more cytokines.*

play a part in the inflammation process, releasing chemicals such as histamines and cytokines in response to physical injury or the presence of a pathogen. A cytokine is a chemical messenger that regulates many different cellular processes to produce a variety of immune responses. Humans have around 40 types of cytokines. In addition to being released from white blood cells, cytokines are also released by the infected cells and induce nearby uninfected cells to release cytokines as well, setting in motion a positive feedback loop that results in a rapid burst of cytokine production.

Cytokines are responsible for triggering the increase in body temperature that causes fever. The higher-than-normal temperatures of a fever inhibit the growth of the invading pathogens and help to speed up the process of repair in the body. One class of cytokines is the interferons. An interferon is a small protein that signals a viral infection to other nearby cells. The interferons stimulate uninfected cells to produce compounds that interfere with viral replication and also activate macrophages and other cells.

The first cytokines and histamines to be produced cause inflammation, a feeling of heat

Any breach of the body's defences against infection triggers a number of responses aimed at combating potential pathogens. One of the more obvious signs of this activity is inflammation or redness at the site of infection.

REACTION OF THE BODY TO A SPLINTER

1 Splinter, redness

2 Chemical signals released by activated mast cells at the injury site cause nearby capillaries to widen and become more permeable

3 Platelets from blood release blood-clotting proteins at wound site

4 Chemical signals emitted by mast cells attract neutrophils from the blood to the site of injury. Neutrophils phagocytose pathogens and tissue heals

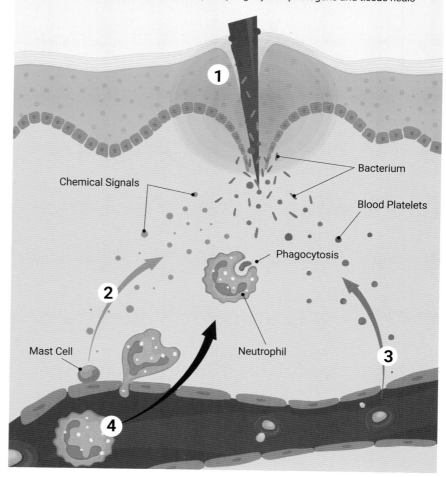

and pain, and swelling and reddening of the skin at the site of the infection or injury. The chemical signals that trigger this response cause the capillary walls to become more permeable, so that serum and other compounds leak out causing swelling which results in pain. This swelling provides a physical barrier that stops the spread of the infection. White blood cells are attracted to the area of inflammation, the particular types depending on the nature of the injury or infecting pathogen. Neutrophils, the most abundant white blood cells of the immune system, are among the first to arrive and will engulf the pathogens. Macrophages follow and will clean up cell debris and pathogens.

A particular type of lymphocyte produced by the lymphatic system called a natural killer cell also plays a role in the innate immune system. Natural killer cells can attack and kill cells infected with viruses which they identify by detecting changes in the proteins on the surface

COMPLEMENT SYSTEM

The complement system gets its name because it is complementary to the innate and adaptive immune systems. It is an array of over 20 different proteins synthesized in the liver. Complement proteins bind to the surfaces of microorganisms, serving as markers to indicate the presence of a pathogen to phagocytic cells. Some complement proteins can combine together in a chain of steps called a complement cascade to open up pores in the cell membranes of invading microbes, causing their death.

of an infected cell. They do this by triggering a process called apoptosis, or cell death. They will also destroy cells which have become cancerous.

THE ADAPTIVE IMMUNE SYSTEM

The adaptive immune system is not as fast acting as the innate immune system. It can take days or even weeks for the adaptive response to become established. However, unlike the non-specific response of the innate immune system, the adaptive system is specifically tailored to dealing with particular pathogens. The adaptive system works in tandem with, and relies on information from, the innate system. It is composed of highly specialized cells and processes that target and eliminate specific pathogens and tumour cells.

An adaptive immune response is set in motion by antigens that the immune system recognizes as foreign. An antigen is a small, specific molecule, such as a sequence of amino acids, on a particular pathogen that stimulates a response in the immune system. Unlike the innate immune response, an adaptive immune response is highly specific to a particular pathogen (or its antigen). There are millions of small molecules that can potentially act as antigens. The adaptive immune system works because the immune cells involved are each able to recognize and respond to one specific antigen, or a few very similar ones.

Immune System Cells

The main cells of the adaptive immune system are lymphocytes, white blood cells that arise in the red bone marrow and mature either there or in the thymus. B cells remain in the bone marrow to mature, while T cells migrate to the thymus to mature there. During the maturation process, each B or T cell develops unique surface proteins that are able to recognize a unique set

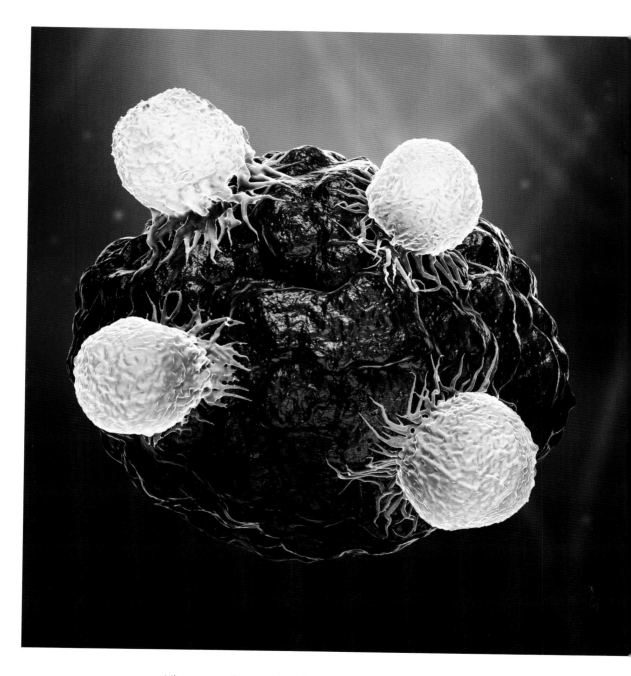

Like a scene from science fiction, a number of killer T cells surround and attempt to destroy a cancer cell.

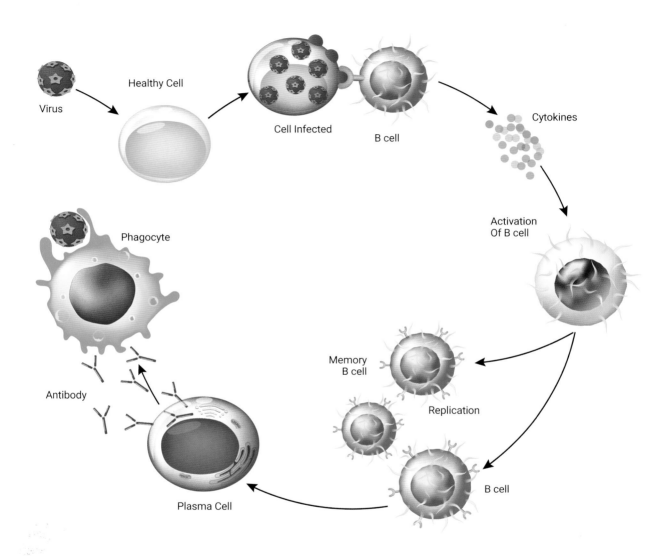

Virus

Healthy Cell

Cell Infected

B cell

Cytokines

Activation Of B cell

Memory B cell

Replication

B cell

Phagocyte

Antibody

Plasma Cell

The infection of a cell by an invading virus triggers a series of events.
Cytokines activate more B cells which divide to form cells which directly
enter the attack and memory cells which are ready to respond against future
infections providing lifelong immunity. Plasma cells, also known as effector B
cells, release antibodies that attract phagocytes to attack the viruses.

of very specific antigens. A single B or T cell that can bind to a tuberculosis bacillus won't be able to bind to an influenza virus, for example.

Individually, each B or T cell can recognize only a very few different molecules, but together the entire lymphocyte population in a healthy person should be able to recognize molecules from most pathogens. At least 10 million different antigen receptors, each with a unique 3D shape, are necessary to recognize all the pathogens an individual may encounter in a lifetime. These unique surface proteins, or receptors, on the lymphocytes are determined genetically – you are born with your arsenal of pathogen detectors. No mature B or T cells will recognize and bind to molecules that are found on healthy human cells but will only respond to molecules found on pathogens or on unhealthy human cells. Exceptions to this occur in the case of autoimmune diseases such as arthritis, lupus and type I diabetes, when the lymphocytes mistakenly attack healthy cells.

B cells and killer T cells are collectively called effector cells because they are involved in bringing about, or effecting, the immune response of killing pathogens and infected host cells. B cells target pathogens in blood and lymph by secreting antibodies. T cells target infected cells in the body. T cells include the helper T cells and the cytotoxic, or killer, T cells which directly kill human cells that are infected or unhealthy. Cytotoxic cells are particularly important in protecting against viral infections. Because viruses replicate within cells they are shielded from contact with antibodies by the infected cell itself. Killer T cells identify and destroy infected cells before the virus can replicate and escape, thereby halting the spread of the infection. Phagocytes from the innate immune system clean up the resulting cellular debris, ingesting and destroying any pathogens that were inside the infected cells.

Helper T cells do not directly kill infected cells, but are crucial for the function of all other cells in the immune response to a pathogen, controlling their response by releasing chemical messengers such as cytokines. Marking out a pathogen by binding antibodies to it is only the first step. The other leukocytes cannot take action against the pathogens they encounter without being activated by the helper T cells. HIV (the human immunodeficiency virus) is such a serious pathogen because it infects and ultimately destroys the helper T cells. If the number of helper T cells declines below a certain level, they cannot activate the other leukocytes to act effectively against other pathogens, so the affected person becomes more susceptible to infections they could otherwise have overcome, resulting in the development of acquired immune deficiency syndrome, or AIDS.

Immunity

An important function of the adaptive immune system that is not shared by the innate immune system is the creation of immunological memory, or immunity. This occurs after the initial response to a specific pathogen and allows for a faster, stronger response if the pathogen in encountered again. The adaptive immune response can be so effective that the pathogen is dealt with before it can even cause symptoms of illness.

Once a pathogen has been eliminated from the body, most of the activated T cells and B cells die within a few days, their purpose served. A few, however, survive and remain in the body as memory T cells or memory B cells, ready to respond immediately if the pathogen is encountered again in the future. This is how we acquire immunity. Active immunity can also be acquired artificially through immunization, the deliberate exposure of a person to a pathogen in order to stimulate the formation of memory cells specific to that pathogen. Typically, only part of the pathogen – a weakened form of the pathogen or a killed pathogen – is used in a vaccine, which triggers an adaptive immune response without causing the full-blown illness.

Passive immunity comes about when pathogen-specific antibodies or activated T cells are transferred to a person who has never been previously exposed to the pathogen. Passive immunity provides immediate protection but doesn't result in the formation of an immunological memory that would give protection from the same pathogen in the future. Immunity lasts only as long as the transferred antibodies or T cells survive in the blood, usually no more than a few months at best.

Passive immunity is acquired naturally when antibodies are transported from mother to foetus across the placenta and may also be acquired by an infant through the mother's breast milk. This provides protection from common pathogens in the environment while the child's own immune system matures. When there is a high risk of infection, older children and adults can acquire artificial passive immunity through the injection of antibodies or activated T cells.

ALLERGY

An allergy occurs when the immune system is hypersensitive to an antigen in the environment that causes little or no response in most people. Any antigen that causes an allergy is called an allergen. Common allergens are plant pollens, insect stings, dust mites, foods such as peanuts or shellfish and some medications, such as aspirin. Allergic symptoms vary depending on the type of exposure, and the severity of the immune system response. At the first exposure to the allergen, B cells are activated, producing large amounts of antibodies which attach to mast cells. The mast cells immediately release cytokines and histamines, which in turn cause inflammation and other responses associated with the signs and symptoms of allergies.

The most severe allergic reaction, caused by a massive release of histamines, is called anaphylaxis. Symptoms of anaphylaxis include falling blood pressure, shortness of breath, and swelling of the tongue and throat, which may be life-threatening. Injecting epinephrine helps control the immune reaction by constricting blood vessels to increase blood pressure, relaxing smooth muscles in the lungs to improve breathing and reducing swelling.

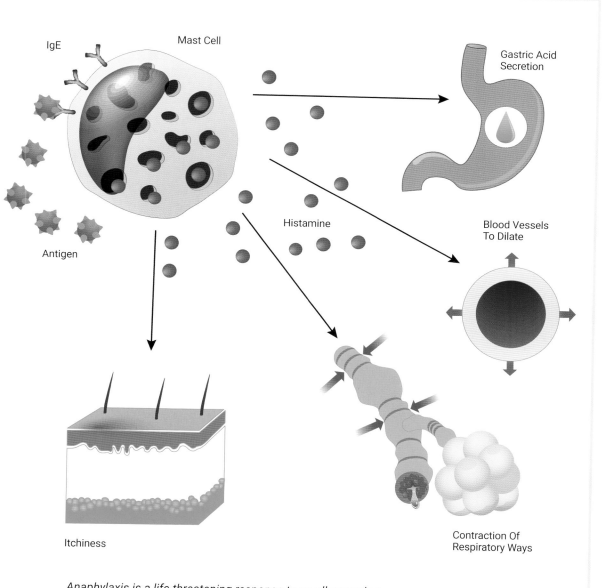

IgE

Mast Cell

Gastric Acid
Secretion

Histamine

Blood Vessels
To Dilate

Antigen

Itchiness

Contraction Of
Respiratory Ways

Anaphylaxis is a life-threatening response to an allergen that results in a massive release of histamines, causing restriction of the airways and dilation of blood vessels among other responses.

CHAPTER THIRTEEN

The Endocrine System

The Endocrine System

Along with the nervous system, the endocrine system is one of the body's major communications networks. While the nervous system is involved in rapid responses to events in the external environment, the endocrine system plays an important role in homeostasis, maintaining the balance of the body's internal environment. The glands of the endocrine system release chemical messenger molecules, called hormones, into the bloodstream. Unlike the rapidly transmitted electrical messages of the nervous system, endocrine hormones act more slowly, travelling through the bloodstream to the cells they affect. Endocrine hormones may affect many cells and have long-lasting, body-wide effects.

The endocrine system involves a complex interplay of actions and reactions between hormones, the glands that produce them and the target organs they affect. Hormones carry instructions from more than a dozen endocrine glands and tissues to cells all over the body. About 50 different known hormones have been identified in humans, controlling a variety of biological processes including muscle growth, heart rate, menstrual cycles and hunger. A single hormone may affect more than one process and one function may be controlled by several hormones.

Endocrine glands are located throughout the body. The primary function of the endocrine gland is to secrete their hormones into the bloodstream which then transports the hormones throughout the body. The endocrine system includes the pituitary, thyroid, parathyroid, adrenal and pineal glands. The hypothalamus, thymus, heart, kidneys, pancreas, stomach, small intestine, liver, skin, female ovaries and male testes also contain cells with endocrine functions.

HORMONES

Hormones play a vital role in the regulation of physiological processes within the body. Although a hormone may be carried throughout the body in the bloodstream, it will only affect its target cells; that is, cells with receptors for that particular hormone. Cells lacking the appropriate receptors will not respond to the hormone. Once the hormone binds to the receptor, a process begins that leads to a response from the target cells. These responses contribute to activities such as reproduction, the growth and development of body tissues, metabolism, sleep and many other body functions.

Hormones can be divided into two major groups based on their chemical structure – steroid and non-steroid hormones. These chemical groups affect a hormone's distribution, the type of receptors it binds to, and other aspects of its function.

A steroid hormone is formed from lipids and is fat soluble, a characteristic that enables it to diffuse across the target cell's plasma membrane, which is also made of lipids. Once inside the cell, a steroid hormone binds with receptor proteins in the cell's cytoplasm to form a structure called a steroid complex. This moves into the cell's nucleus, where it influences the expression of genes. Examples of steroid hormones include cortisol, which is secreted by the adrenal glands, and sex hormones, such as oestrogen, which are secreted by the gonads.

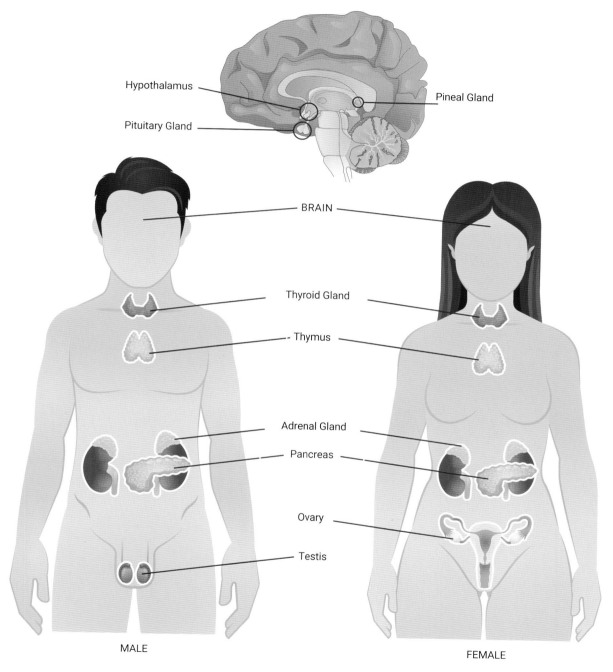

Hypothalamus

Pineal Gland

Pituitary Gland

BRAIN

Thyroid Gland

- Thymus

Adrenal Gland

Pancreas

Ovary

Testis

MALE

FEMALE

The locations in the body of the major components of the endocrine system.

Steroid hormones can pass directly through the membrane of a cell, setting in train a series of events that results in the cell producing a particular protein.

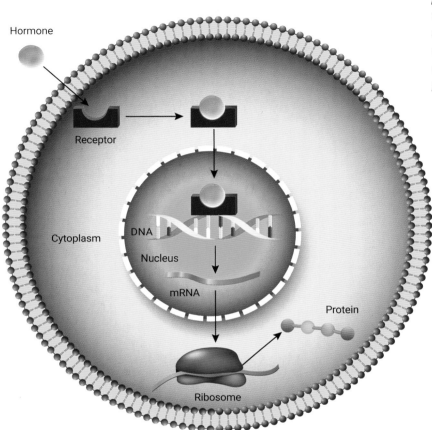

Hormone

Receptor

Cytoplasm

DNA

Nucleus

mRNA

Protein

Ribosome

A non-steroid hormone is formed from amino acids. Because it is not fat soluble, it cannot diffuse across the plasma membrane of a target cell in the way a steroid hormone can. Instead, it binds to a receptor protein on the cell membrane and in doing so activates an enzyme in the membrane. The enzyme then stimulates another molecule, called the second messenger, which influences processes inside the cell. Examples of non-steroid hormones include glucagon and insulin, both produced by the pancreas and responsible for

regulating blood sugar levels. Most endocrine hormones are non-steroid hormones.

Feedback Loops

The production of hormones in the body is regulated by feedback loops, which may be either negative or positive. Most hormones are regulated by negative feedback loops which keep the concentration of a hormone within a relatively narrow range and maintain homeostasis. For example, the secretion of hormones by the

Non-steroid hormones cannot enter cells directly, but must instead bind to a receptor protein which activates an enzyme which in turn stimulates a so-called 'second messenger' which triggers processes inside the cell.

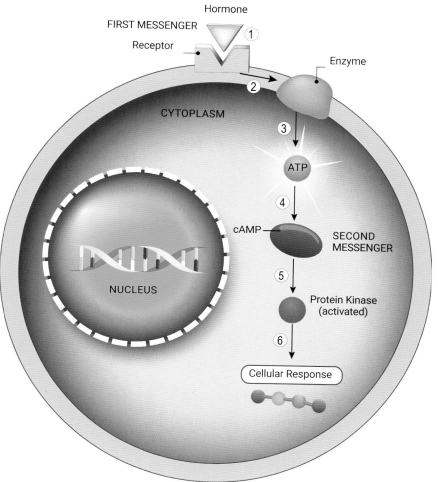

thyroid gland is controlled by a negative feedback loop that involves two other glands. When thyroid hormone levels in the blood fall too low, the hypothalamus secretes thyrotropin-releasing hormone (TRH) which travels directly to the pituitary gland where it stimulates the pituitary to secrete thyroid-stimulating hormone (TSH). TSH travels through the bloodstream to the thyroid gland, stimulating it to secrete thyroid hormones. When levels of thyroid hormones in the blood are high enough once more, this feeds back to the

hypothalamus, stopping it from secreting TRH and subsequently the pituitary from secreting TSH, and consequently the thyroid gland stops secreting its hormones. When, eventually, the levels of thyroid hormones in the blood fall too low again, the hypothalamus releases TRH, and the loop repeats.

Very few endocrine hormones are regulated by positive feedback loops, which cause the concentration of a hormone to become increasingly higher. An example of this is the

Negative feedback loops are an important mechanism in maintaining a homeostatic balance in the body's internal environment. In this example, the level in the blood of glucocorticoids, hormones which influence the metabolism of proteins, fat and sugars, is regulated by the release of corticotropin-releasing hormone (CRH) from the hypothalamus.

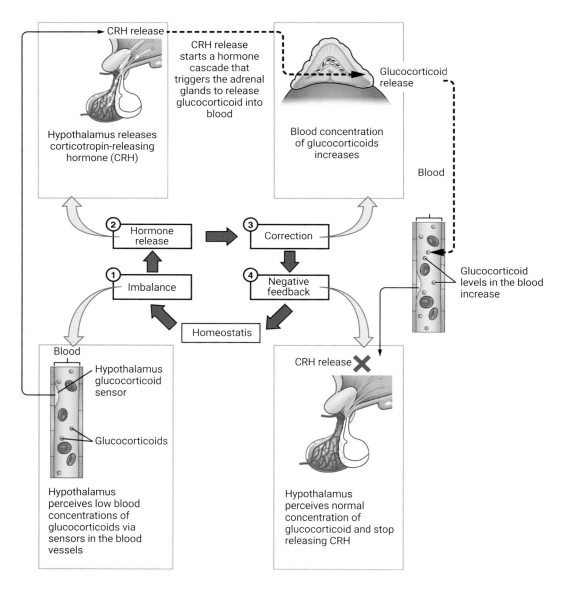

CRH release

CRH release starts a hormone cascade that triggers the adrenal glands to release glucocorticoid into blood

Hypothalamus releases corticotropin-releasing hormone (CRH)

Glucocorticoid release

Blood concentration of glucocorticoids increases

Blood

② Hormone release

③ Correction

① Imbalance

④ Negative feedback

Homeostatis

Glucocorticoid levels in the blood increase

Blood

Hypothalamus glucocorticoid sensor

Glucocorticoids

Hypothalamus perceives low blood concentrations of glucocorticoids via sensors in the blood vessels

CRH release ✕

Hypothalamus perceives normal concentration of glucocorticoid and stop releasing CRH

production of prolactin, which is secreted by the pituitary gland. When a baby suckles on the mother's nipple, nerve impulses from the nipple reach the hypothalamus, which stimulates the pituitary gland to secrete prolactin. Prolactin travels in the blood to the mammary glands and stimulates them to produce milk so the baby continues to suckle, causing more prolactin to be secreted and more milk to be produced. The positive feedback loop continues until the baby stops suckling at the breast.

THE PITUITARY GLAND

The pituitary gland is about the size of a pea and protrudes from the bottom of the hypothalamus at the base of the brain. A thin stalk, called the infundibulum, connects the pituitary to the hypothalamus and allows direct connections

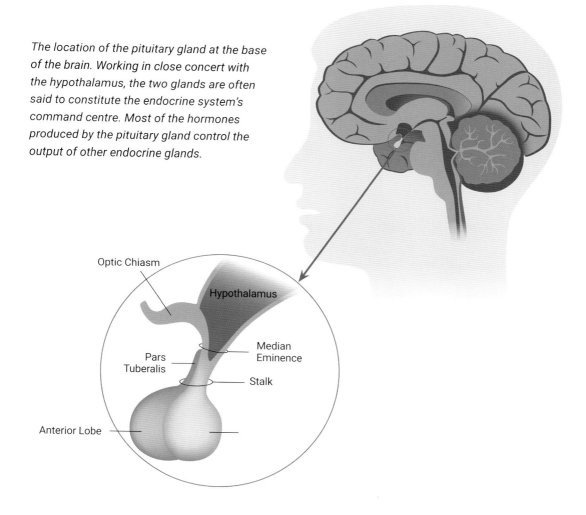

The location of the pituitary gland at the base of the brain. Working in close concert with the hypothalamus, the two glands are often said to constitute the endocrine system's command centre. Most of the hormones produced by the pituitary gland control the output of other endocrine glands.

between the two structures. Together, the hypothalamus and pituitary glands are often described as the 'command centre' of the endocrine system.

The pituitary gland has two lobes. The anterior pituitary is the lobe at the front of the pituitary gland. It is regulated mainly by hormones from the hypothalamus which secretes hormones called releasing hormones and inhibiting hormones that travel directly to the anterior lobe, stimulating it to either release or stop releasing particular pituitary hormones. The posterior pituitary at the back of the pituitary gland does not synthesize any hormones, but instead stores hormones from the hypothalamus along the axons of nerves connecting the two structures, secreting the hormones into the bloodstream as needed. Hypothalamic hormones secreted by the posterior pituitary include vasopressin, or antidiuretic hormone, which stimulates the kidneys to conserve water by producing more concentrated urine.

THE THYROID GLAND

The thyroid gland, located in the front of the neck below the trachea, or Adam's apple, is one of the largest of the endocrine glands. It is butterfly shaped and has two lobes connected by a narrow band of tissue called an isthmus. Internally, the thyroid gland is composed mainly of follicles, small clusters of cells surrounding a central cavity, which store hormones and other molecules. The outer layer of cells of each follicle secretes thyroid hormones as needed. The cells of the thyroid follicles are highly specialized to absorb iodine, in the form of iodide ions, which they use to produce thyroid hormones. The cells also use some of the iodide they absorb to form a protein called thyroglobulin, which stores iodide

for later hormone synthesis. Scattered among the follicles are parafollicular cells, which synthesize and secrete the hormone calcitonin.

Calcitonin helps to regulate blood calcium levels by stimulating the absorption of calcium into bone tissue. Calcitonin works in parallel with parathyroid hormone, which is secreted by the parathyroid glands and has the opposite effect, causing calcium to be redeposited into the bloodstream.

The main thyroid hormones are produced by the follicles: triiodothyronine (T3) and thyroxine (T4). T4 makes up about 90 per cent of thyroid hormone, but T3 is much more powerful. However, most of the T4 is converted to T3 by target tissues.

Both T3 and T4 cross cell membranes with the help of transporter proteins. Inside the nucleus of cells, T3 and T4 activate genes that control protein synthesis, increasing the rate of metabolism in cells, so they absorb more carbohydrates, use more energy and produce more heat. Thyroid hormones also increase the rate and force of the heartbeat. In addition, they increase the sensitivity of cells to fight-or-flight hormones, such as adrenaline.

The production of both T3 and T4 is regulated primarily by thyroid-stimulating hormone (TSH), which is secreted by the anterior pituitary gland. TSH production, in turn, is regulated by thyrotropin-releasing hormone (TRH), which is produced by the hypothalamus. The thyroid gland, pituitary gland and hypothalamus form a negative feedback loop to keep thyroid hormone secretion within a normal range (see page 215).

ADRENAL GLANDS

The adrenal glands are located on both sides of the body, just above the kidneys. The right adrenal

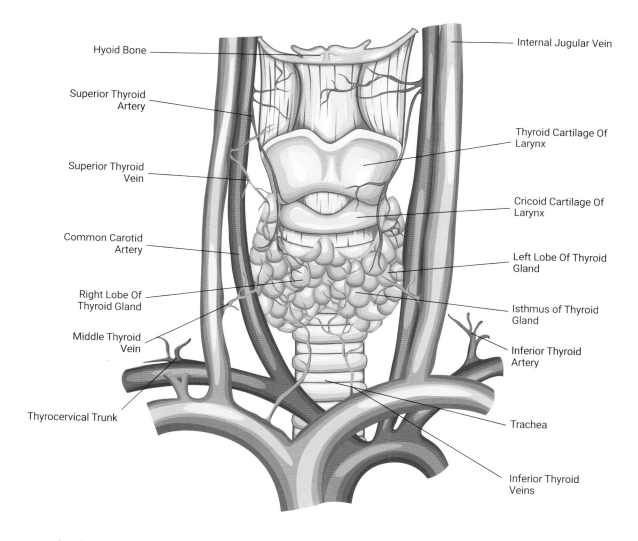

Hyoid Bone

Superior Thyroid Artery

Superior Thyroid Vein

Common Carotid Artery

Right Lobe Of Thyroid Gland

Middle Thyroid Vein

Thyrocervical Trunk

Internal Jugular Vein

Thyroid Cartilage Of Larynx

Cricoid Cartilage Of Larynx

Left Lobe Of Thyroid Gland

Isthmus of Thyroid Gland

Inferior Thyroid Artery

Trachea

Inferior Thyroid Veins

gland is smaller and has a pyramidal shape. The larger, left adrenal gland has a half-moon shape. Each adrenal gland has two distinct parts. An outer layer, called the adrenal cortex, produces steroid hormones, such as cortisol, while the inner layer, called the adrenal medulla, produces non-steroid hormones, such as adrenaline.

The adrenal cortex is subdivided into three layers, called zones. Each zone has distinct

Located in the front of the neck below the trachea, the thyroid produces hormones which increase the rate of metabolism in cells throughout the body.

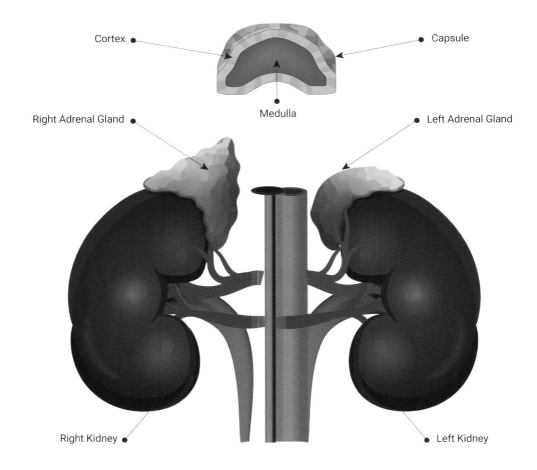

Sitting above the kidneys, the adrenal glands secrete a number of different hormones including adrenaline, involved in the fight-or-flight response, and cortisol, which is considered to be the primary stress hormone.

enzymes that produce different hormones. Hormones produced by the adrenal cortex are known by the general term corticosteroids. Three types of corticosteroids are synthesized and secreted by the adrenal cortex, each type produced by a different zone of the cortex.

Mineralocorticoids are produced in the outermost zone, called the zona glomerulosa. These hormones help control the balance of mineral salts (electrolytes) in the body, which affects blood pressure. Aldosterone increases the reabsorption of sodium ions and the excretion of potassium ions in the kidneys and also stimulates the retention of sodium ions by cells in the colon.

Glucocorticoids are produced in the middle layer, called the zona fasciculata. Glucocorticoids help control the rate of metabolism of proteins, fats and sugars and also have a potent anti-inflammatory effect. One of the main glucocorticoids is cortisol, which is released in response to stress. Cortisol also reduces the production of new bone and decreases absorption of calcium from the gastrointestinal tract.

Androgens are produced in the inner zone, the zona reticularis. Although androgen is used as a general term for male sex hormones, this is somewhat misleading, as adrenal cortex androgens are produced by both males and females. In adult males, they are converted to more potent androgens, such as testosterone in the male testes, while in adult females, they are converted to oestrogens in the female ovaries.

Similar to the thyroid gland, hormone production in the adrenal cortex is regulated by a negative feedback loop involving the pituitary gland and the hypothalamus. For example, corticotropin-releasing hormone (CRH) from the hypothalamus stimulates production of adrenocorticotropic hormone (ACTH) from the anterior pituitary, which in turn stimulates

The adrenal glands are divided into distinct areas, each producing a particular kind of hormone. Corticosteroids are produced in the zones of the adrenal cortex; mineralocorticoids in the zona glomerulosa; glucocorticoids in the zona fasciculata, and androgens in the zona reticularis. Non-steroid hormones such as adrenaline are produced in the adrenal medulla.

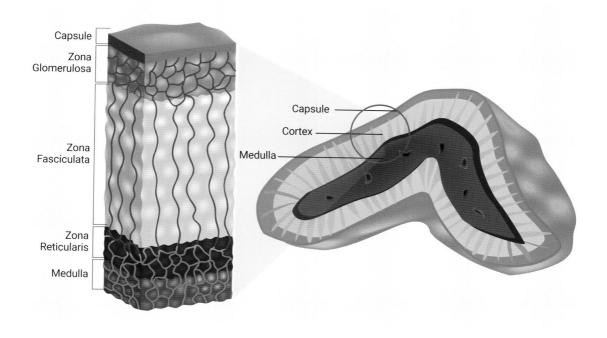

production of cortisol by the adrenal cortex. When levels of glucocorticoids like cortisol start to rise too high, the hypothalamus and pituitary gland stop secreting CRH and ACTH, respectively. When levels of glucocorticoids start to fall too low the opposite occurs.

The hormones synthesized by the adrenal medulla at the centre of the adrenal gland are secreted into a dense network of blood vessels. Generally known as catecholamines, they include adrenaline (or epinephrine) and noradrenaline (or norepinephrine). These are water-soluble, non-steroid hormones formed of amino acids. Catecholamines produce a rapid response throughout the body in stressful situations, bringing about changes such as increased heart

The pancreas fulfils a dual role as part of both the digestive and endocrine systems. Acinar cells secrete digestive enzymes which pass through ducts into the small intestine. Hormones are secreted into a network of capillaries by clusters of cells called islets of Langerhans. Different types of cell in the islets secrete different hormones, all of which are involved in the regulation of glucose levels in the blood.

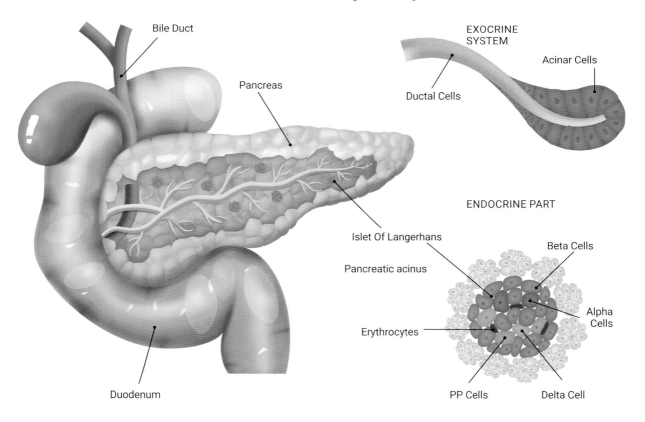

rate, rapid breathing, an increase in blood pressure and the constriction of blood vessels in certain parts of the body. The release of catecholamines by the adrenal medulla is stimulated by activation of the sympathetic division of the autonomic nervous system (see page III).

THE PANCREAS

The pancreas, a large gland located in the upper left abdomen behind the stomach, serves as both an endocrine gland and an exocrine gland. The difference between endocrine and exocrine glands is that endocrine glands, such as the adrenal glands, secrete their products directly into the bloodstream, whereas exocrine glands secrete substances via ducts on to an epithelial surface, for example sebaceous glands in the skin. It is about 15 cm (6 in) long, and is roughly a flat, oblong shape. As well as being part of the endocrine system, the pancreas is also part of the digestive system, releasing digestive enzymes via ducts into the gastrointestinal tract.

Clusters of cells within the pancreas called the islets of Langerhans, or pancreatic islets, secrete endocrine hormones into a dense network of capillaries. There are approximately three million of these pancreatic islets which consist of four main types of cells, each of which secretes a different endocrine hormone. All of the hormones produced by the pancreatic islets play crucial roles in glucose metabolism and the regulation of blood glucose levels.

Islet cells called alpha (α) cells secrete the hormone glucagon, which stimulates the liver to convert stored glycogen into glucose, which is released into the bloodstream. Beta (β) cells secrete insulin which stimulates the absorption of glucose from the blood into fat, liver and skeletal muscle cells where it is converted into glycogen,

DIABETES

The most common type of pancreatic disorder is diabetes mellitus. There are two major types, type 1 diabetes and type 2 diabetes, each with different causes, but generally with the same symptoms. Type 1 diabetes is a chronic autoimmune disorder in which the immune system attacks the insulin-secreting beta cells of the pancreas. As a result, people with type 1 diabetes lack the insulin needed to keep blood glucose levels within the normal range. For type 1 diabetics, insulin injections are critical for survival.

Type 2 diabetes is the most common form of diabetes and usually includes a combination of insulin resistance along with impaired insulin secretion, resulting in high blood glucose levels. Both genetic and environmental factors play a role in the development of type 2 diabetes. Type 2 diabetes can be managed with changes in diet and physical activity along with medication.

fats (triglycerides), or both. Together glucagon and insulin maintain blood sugar levels.

Delta (δ) cells secrete somatostatin, also called growth hormone inhibiting hormone, because it inhibits the anterior lobe of the pituitary gland from producing growth hormone. Somatostatin also inhibits the secretion of pancreatic endocrine hormones and pancreatic exocrine enzymes. Gamma (γ) cells secrete pancreatic polypeptide which helps to regulate the secretion of both endocrine and exocrine substances by the pancreas.

The Reproductive System

The Reproductive System

The reproductive system is the human organ system that is responsible for the continuation of the species, achieved through the production and fertilization of gametes (sperm and eggs). The reproductive system is the only organ system that differs significantly between males and females.

In general, the reproductive structures in humans can be divided into three main categories: gonads, internal genitalia and external genitalia. The gonads are the organs in which gametes, the cells that fuse in fertilization to form new individuals, develop and mature – all other reproductive structures are termed

The organs of the male reproductive system. Most are located outside the body.

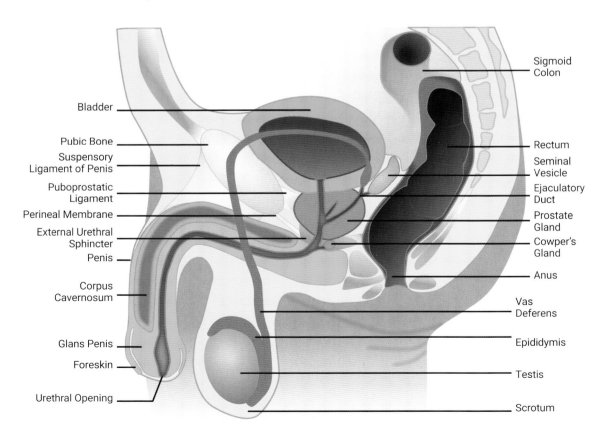

Bladder

Pubic Bone

Suspensory Ligament of Penis

Puboprostatic Ligament

Perineal Membrane

External Urethral Sphincter

Penis

Corpus Cavernosum

Glans Penis

Foreskin

Urethral Opening

Sigmoid Colon

Rectum

Seminal Vesicle

Ejaculatory Duct

Prostate Gland

Cowper's Gland

Anus

Vas Deferens

Epididymis

Testis

Scrotum

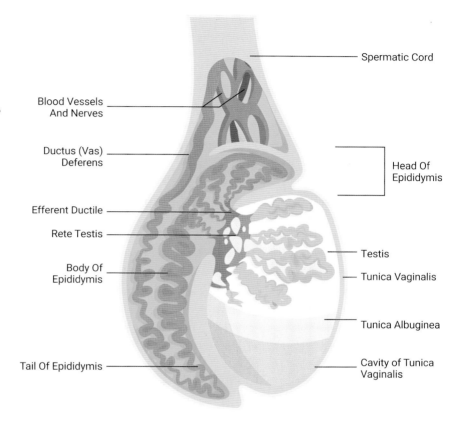

Cross-sectional diagram of a testicle. Sperm travel from the testis, through the epididymis and into the vas deferens.

Spermatic Cord

Blood Vessels And Nerves

Ductus (Vas) Deferens

Head Of Epididymis

Efferent Ductile

Rete Testis

Body Of Epididymis

Testis

Tunica Vaginalis

Tunica Albuginea

Tail Of Epididymis

Cavity of Tunica Vaginalis

genitalia, or genitals. As the terms suggest, internal genitalia are located inside the body, whereas external genitalia are visible on the outside. There are many similarities in both structure and function between males and females. Here, we simply define 'male' and 'female' based on individuals who have the most typical features characteristic of those two sexes while acknowledging that other types of structures are also normal and common. We describe the functions of the sex organs during vaginal sexual intercourse because we are concerned here with the mechanisms of reproduction, but also acknowledge that there are other types of sexual activity which are also common and normal.

THE MALE REPRODUCTIVE SYSTEM

In the male reproductive system, the gonads are the two testicles or testes (singular: testis), each approximately 2.5 by 4 cm (1 x 1.5 in) in size which are contained within the scrotum, a pouch made of skin and smooth muscle that hangs down behind the penis. The testes produce sperm (the male gametes) and some reproductive hormones such as testosterone.

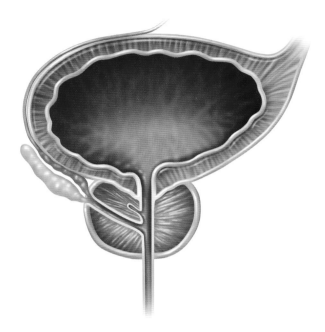

The walnut-sized prostate gland surrounds the urethra where it joins with the ejaculatory ducts that connect the vas deferens with the urethra.

contains a total of about 30 m (100 ft) of these tubules. The tubules contain sperm in several different stages of development. The walls of the seminiferous tubules are made up of the developing sperm cells, with the least developed sperm at the periphery of the tubule and the fully developed sperm in the lumen. The seminiferous tubules within each testis join together to form the efferent ducts which transport immature sperm to the epididymis associated with that testis. The epididymis is a tightly coiled tubule with a total length of about 6 m (20 ft) that resembles a comma and lies along the top and posterior portion of the testes. The epididymis is where the sperm mature and are stored.

The vas deferens, or sperm duct, is a thin tube, about 30 cm (12 in) long, which begins at the epididymis in the scrotum, and continues up into the pelvic cavity. It is made up of ciliated epithelium and smooth muscle. The role of the vas deferens is to transport sperm from the epididymis to the ejaculatory ducts.

Semen is a mixture of sperm and spermatic duct secretions (about 10 per cent of the total) and fluids from accessory glands that contribute most of the semen's volume. An ejaculate (a single emission of sperm) will contain from 2 to 5 millilitres of fluid with 50–120 million sperm per millilitre.

The seminal vesicles are a pair of exocrine glands, each consisting of a single tube, which is folded and coiled upon itself. Each vesicle is about 5 cm (2 in) long and has an excretory duct that merges with the vas deferens to form one of the two ejaculatory ducts, which connect the vas deferens with the urethra. Fluid secreted by the seminal vesicles into the ducts makes up about 60 per cent of the total volume of semen,

Each testis is protected by a multi-layered covering called the tunica. The innermost layer, the tunica vasculosa, consists of connective tissue and contains the blood vessels that supply the testis. The middle layer, the tunica albuginea, is a dense layer of fibrous tissue. The outermost layer, the tunica vaginalis, is actually two layers of tissue separated by a thin fluid layer that helps to reduce friction between the testes and the scrotum.

Inside, the testis is filled with hundreds of tiny, coiled and tightly packed tubes called seminiferous tubules. A single testis normally which is the sperm-containing fluid that leaves the penis during an ejaculation. The fluid from the seminal vesicles contains proteins, fructose and other substances that help nourish sperm; it is also alkaline, so it helps to prolong the lifespan of sperm after it enters the acidic environment inside the female vagina.

The walnut-shaped prostate gland, located just below the seminal vesicles, surrounds the urethra and has a number of short ducts that directly connect to it. It is composed of glandular tissue and smooth muscle, which provides much of the force needed for ejaculation to occur. The glandular tissue makes a thin, milky fluid that contains nutrients, including a high concentration of zinc, which is important for maintaining sperm quality, and enzymes, including a prostate-specific antigen which helps to liquefy the ejaculate several minutes after release from the male. Prostate gland secretions account for about 30 per cent of the bulk of semen.

The pea-sized bulbourethral glands, or Cowper's glands, are located just below the prostate gland. They release a protein-rich alkaline fluid that accounts for a couple of drops of fluid in the total ejaculate. The function of the bulbourethral secretions is to help lubricate the urethra and to neutralize any acid residue in the urethra left over from urine.

The urethra passes through the penis to carry semen from the ejaculatory ducts through the penis and out of the body. After leaving the urinary bladder, the urethra passes through the prostate gland, where the urethra is joined by the ejaculatory ducts. From there, the urethra passes through the penis to its external opening at the tip of the glans penis. Called the external urethral orifice, this opening provides a way for semen, or urine from the bladder, to leave the body.

The Penis

The part of the penis that is located inside the body is called the root of the penis. The shaft of the penis is the part of the penis that is outside the body. The enlarged, bulbous end of the shaft is the glans penis.

The penis is covered with skin that is able to move freely over the body of the penis. In an uncircumcised male, the glans penis is also mainly covered by epithelium, called the foreskin, which is attached to the penis at an area on the underside of the penis called the frenulum.

The reproductive function of the penis is to deliver sperm to the female reproductive tract. This function is called intromission. Intromission depends on the ability of the penis to become stiff and erect, a state referred to as an erection. Unlike many other mammals, human males have no erectile bone in the penis. The penis contains three tubes of erectile tissue that run the length of the organ. These consist of a pair of tubes at the back of the penis, called the corpus cavernosum, and a single tube of tissue at the front, called the corpus spongiosum. The glans penis also contains spongy erectile tissue. During sexual arousal, arteries supplying blood to the penis dilate and these tissues become engorged with blood, becoming erect and hard, in preparation for sexual intercourse. The engorged spongy tissue presses against and constricts the veins that would carry blood away from the penis. The organ is inserted into the vagina, culminating with an ejaculation. During intercourse, the smooth muscle sphincters at the opening to the renal bladder close to prevent urine from entering the penis. An orgasm is a two-stage process: first, glands and accessory organs connected to the testes contract, then semen (containing sperm) is expelled through

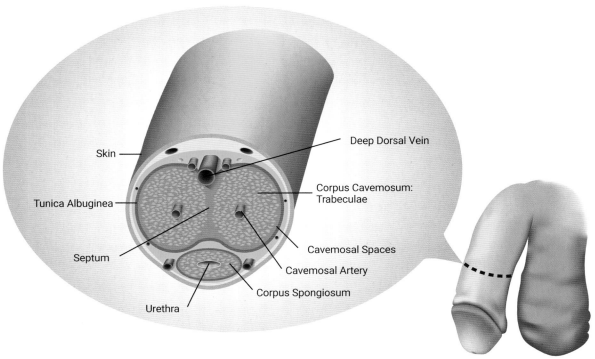

A cross-section through the penis. Internally, the penis consists mostly of spongy tissue that can fill with blood, stiffening the penis during erection.

the urethra during ejaculation. After intercourse, the blood drains from the erectile tissue and the penis becomes flaccid once more.

Sperm

One of the main functions of the male reproductive system is the production of sperm. The process of producing sperm is called spermatogenesis and normally starts when a male reaches puberty. A young, healthy male may produce hundreds of millions of sperm a day and will continue to do so throughout adult life, although production will decline in later years.

Sperm production is very sensitive to temperature, which explains why the testes are located outside the body in the scrotum. A temperature generally about 2 degrees Celsius cooler than core body temperature is optimal for spermatogenesis. The scrotum regulates its internal temperature as needed by contractions of the smooth muscles that line it. If the temperature inside the scrotum becomes too low, the scrotal muscles contract, pulling the scrotum higher against the body, where it is warmer. The opposite occurs if the temperature inside the scrotum becomes too high.

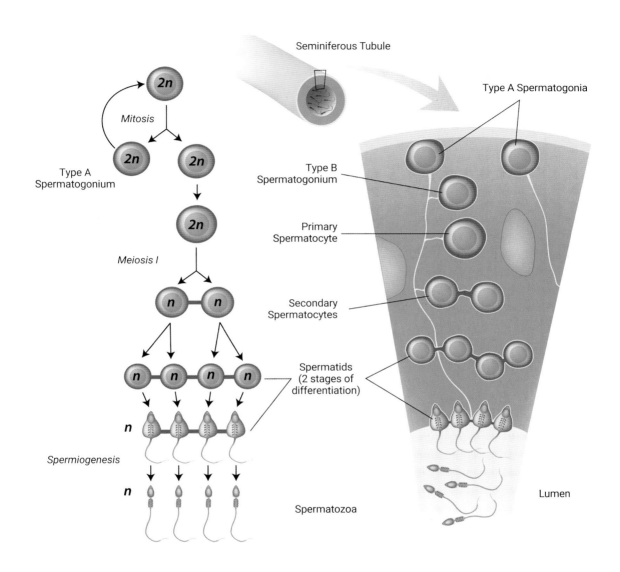

Seminiferous Tubule

Type A Spermatogonia

Type A Spermatogonium

Mitosis

Type B Spermatogonium

Meiosis I

Primary Spermatocyte

Secondary Spermatocytes

Spermatids (2 stages of differentiation)

Spermiogenesis

Spermatozoa

Lumen

Spermatogenesis, the production of sperm, is a multi-stage process that begins nears the walls of the seminiferous tubules and ends with mature sperm in the epididymis. During the process, a diploid spermatogonium gives rise to four haploid spermatids which develop into the mature sperm.

Spermatogenesis takes place in the semi-niferous tubules in the testes. Spermatogenesis requires high concentrations of testosterone, which is produced and secreted by Leydig cells, found between the seminiferous tubules in the testes. 'Nursemaid' cells called Sertoli cells in the tubules have several functions in spermatogenesis. They secrete hormones that help regulate the process; produce substances that initiate meiosis (see below); concentrate testosterone produced by the Leydig cells; and consume the surplus cell material that is discarded from the developing sperm cells.

Chromosomes

The genetic material within a human cell is organized into chromosomes. A human cell contains 23 pairs of these chromosomes. The number of sets of chromosomes in a cell is called its ploidy level. Cells containing two sets

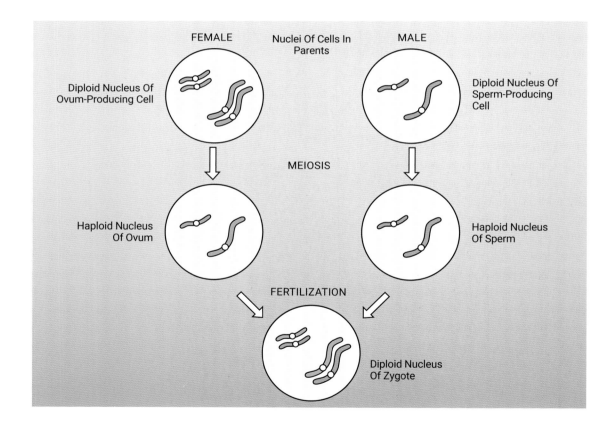

A normal diploid human body cell contains 23 pairs of chromosomes. Sex cells, ova and sperm, are haploid, having just one set of chromosomes. When fertilization takes place, a set of chromosomes from the female combine with a set from the male, giving the full diploid complement of chromosomes.

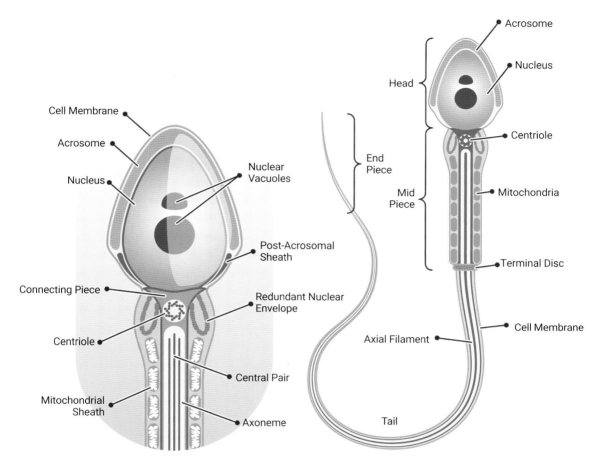

The internal structure of a sperm cell. The many mitochondria in the mid section provide the energy needed to beat the tail filament and propel the sperm cell to its destination.

of chromosomes are termed diploid. Sexual reproduction requires fertilization, a union of two cells from two individual organisms. The diploid cell has to reduce its number of chromosome sets before fertilization can occur otherwise there would be a continual doubling in the number of chromosome sets in every generation. Cells containing one set of chromosomes are termed haploid. The reduction from a diploid to a haploid cell occurs through a process of cell division called meiosis.

The process of spermatogenesis, which takes around ten weeks to complete, begins with a diploid stem cell called a spermatogonium. A spermatogonium divides to produce two cells called primary spermatocytes; one of this pair goes on to produce sperm, while the other adds to the stock of spermatogonia. The primary spermatocyte divides again to produce two haploid daughter cells called secondary spermatocytes which divide again to produce four daughter cells, also haploid, called spermatids.

The spermatids begin to form a tail, their DNA becomes highly condensed, and unnecessary cytoplasm and organelles are removed. The resulting cells are sperm, or spermatozoa, each one a haploid cell, carrying the male parent's contribution to the genetic makeup of any embryo that forms. Each spermatogonium thus gives rise to four sperm cells. A mature sperm cell has several adaptations that help it reach and penetrate an egg. The front of the head is an area called the acrosome, which contains enzymes that will help the sperm penetrate an ovum if it

reaches one. The midpiece, the part of the sperm between the head and the flagellum, contains a great many mitochondria that produce the energy needed to move the flagellum. The flagellum, or 'tail', rotates like a propeller, providing the propulsion for the sperm to 'swim' through the female reproductive tract to reach an ovum.

Although the sperm produced in the testes have flagella, they do not yet have the ability to move. They acquire this in the epididymis, into which they are transported in testicular fluid secreted by the Sertoli cells. The mature sperm are stored in the epididymis until ejaculation occurs.

Ejaculation happens when peristaltic contractions of the muscle layers of the vas deferens and other structures propel sperm from

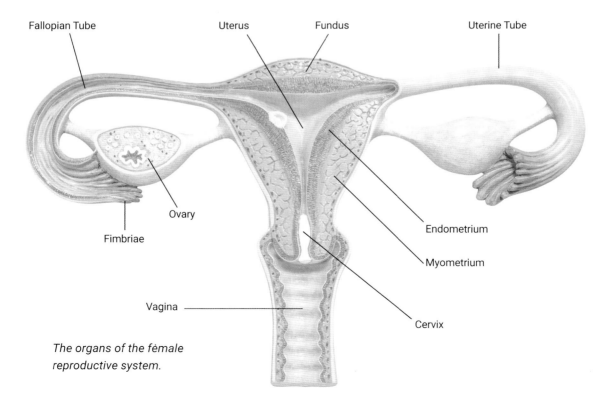

The organs of the female reproductive system.

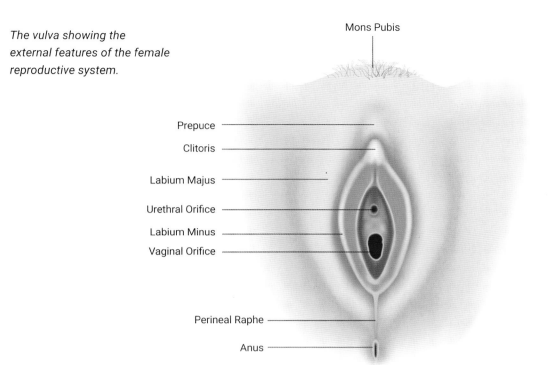

The vulva showing the external features of the female reproductive system.

Mons Pubis

Prepuce

Clitoris

Labium Majus

Urethral Orifice

Labium Minus

Vaginal Orifice

Perineal Raphe

Anus

the epididymis. The muscle contractions force the sperm through the vas deferens and the ejaculatory ducts, and out of the penis through the urethra in a series of spurts. As the sperm travel through the ejaculatory ducts, they mix with secretions from the seminal vesicles, prostate gland and bulbourethral glands to form semen. Hundreds of millions of mature sperm are propelled from the penis during a normal ejaculation.

FEMALE REPRODUCTIVE ORGANS

The internal and external organs of the female reproductive system function to produce female gametes called ova (or oocytes), secrete female sex hormones (such as oestrogen), and carry and give birth to a foetus. The internal female reproductive organs include the vagina, uterus, oviducts and ovaries. The external organs include

the breasts and the vulva, which consists of the mons pubis, clitoris, labia majora, labia minora and the vestibular glands.

The vulva is the name for the entire set of external genitalia in the groin area of females; although commonly this is sometimes referred to as the vagina, anatomically speaking the vagina is an entirely internal structure.

The mons pubis is a pad of fat that overlies the pubic bone. The labia majora (labia = lips; major = larger) are a pair of elongated folds of tissue that begin just to the rear of the mons pubis and enclose the other components of the vulva. In the developing embryo, the labia majora derive from the same tissue that produces the scrotum in a male. The labia minora are thin folds of tissue centrally located within the labia majora. These labia are more pigmented and protect

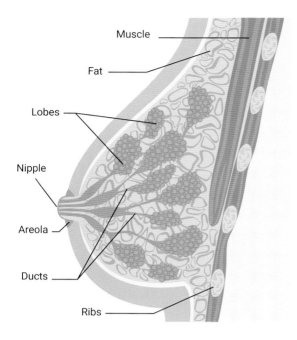

Muscle

Fat

Lobes

Nipple

Areola

Ducts

Ribs

The anatomy of a breast. Milk is secreted by the lobes of the mammary glands and drains into the nipple.

the openings to the vagina and urethra. The mons pubis and the front part of the labia majora become covered with hair during adolescence; the labia minora is hairless.

The upper parts of the labia minora come together to encircle the clitoris, an organ that in the embryo develops from the same cells as the glans penis in the male. It is a structure with erectile tissue that contains a large number of sensory nerves and serves as a source of stimulation during intercourse.

The hymen is a thin membrane that sometimes partially covers the entrance to the vagina. An intact hymen is often taken to indicate virginity but it is only ever a partial membrane, as

menstrual fluid and other secretions must be able to exit the body, regardless of whether intercourse has ever taken place. The opening of the vagina is flanked by the greater vestibular glands or Bartholin's glands, which provide lubrication during intercourse.

The Breasts

Although they are located far from the other organs, breasts are generally considered as part of the female reproductive system. The function of the breasts is to supply milk to nourish an infant in a process called lactation. Multiple bands of connective tissue called suspensory ligaments support and connect the breast tissue to the dermis of the overlying skin.

The external features of the breast include a nipple surrounded by a generally circular areola, varying in size from 2.5 to 10 cm (1–4 in) or so, the colour of which may deepen during pregnancy. The areolar region is characterized by small, raised areolar glands that secrete lubricating fluid during lactation to protect the nipple from chafing. Breast milk is produced by the mammary glands, which are modified sweat glands, in groups of milk-secreting cells in clusters called alveoli. Fat tissue surrounding the glandular lobes determines the size of the breast, which differs between individuals and does not affect the amount of milk produced. Breast tissue responds to changing levels of oestrogen and progesterone through the menstrual cycle which can lead to swelling and breast tenderness in some individuals. If pregnancy occurs, the increase in hormones leads to further development of the mammary tissue and enlargement of the breasts.

Once milk is made in the alveoli, myoepithelial cells surrounding the alveoli contract to push the milk to the lactiferous sinuses and then to the 15 to 20 lactiferous ducts that open on the surface of the nipple, from which the baby can draw milk by suckling. When a baby nurses from the breast, the entire areolar region is taken into the mouth.

The Vagina

In the context of pregnancy and natural (vaginal) childbirth, the vagina is often referred to as the birth canal. It is the entrance to the reproductive tract and channels the flow of menstrual blood from the uterus. The vagina is supported within the pelvic cavity by muscles and ligaments. It is an elastic, muscular canal about 10 cm (4 in) long leading from its opening in the vulva to the neck of the uterus, made up of layers of muscle and fibrous tissue lined with mucous membranes. Folds in the mucosa allow the vagina to stretch to many times its normal size during birth. The vagina accommodates the penis and is the site where sperm are usually ejaculated during sexual intercourse.

THE UTERUS

The uterus (commonly called the womb) is the muscular pear-shaped organ that nourishes and supports the growing embryo. It is located above the vagina and behind the bladder in the centre of the pelvis, held in place by several ligaments and bands of supportive tissue. In the non-pregnant female, its average size is approximately 5 cm (2 in) wide by 7 cm (2.75 in) long. It has three sections. The cervix is the narrow portion of the uterus that projects into the vagina. The middle section of the uterus is called the body of the uterus. The portion of the uterus where the uterine tubes enter is called the fundus.

The wall of the uterus is made up of three layers. The outermost layer is the serous membrane, or perimetrium, formed of epithelial tissue. The middle layer, the myometrium, is a thick layer of smooth muscle responsible for contractions of the uterus. Most of the uterus is formed from this tissue, with muscle fibres running horizontally, vertically and diagonally, allowing the powerful contractions that occur during labour as well as the less powerful contractions (or cramps) that occur when menstrual blood is expelled during a woman's period.

The innermost layer of the uterus is called the endometrium, consisting of a connective tissue lining, called the lamina propria, covered by epithelial tissue lining the inside of the uterus. Part of the endometrium grows and thickens in response to increased levels of the hormones oestrogen and progesterone, providing a site of implantation for a fertilized egg. Should fertilization not take place this layer of the endometrium sheds during menstruation.

The Ovaries

The ovaries are the female gonads. They are each about 2 to 3 cm (1 in) in length, about the size and shape of an almond. The ovaries are located within the pelvic cavity, and are supported by the mesovarium, an extension of the peritoneum. The suspensory ligament, containing the ovarian blood and lymph vessels extends from the mesovarium, and the ovary itself is attached to the uterus via the ovarian ligament.

The ovary has an outer covering called the ovarian surface epithelium over a dense covering of connective tissue beneath which is the cortex, or outer portion, of the ovary. The cortex is composed of a tissue framework called the ovarian stroma, within which the oocytes develop surrounded by supporting cells. This grouping of an oocyte and its supporting cells is called a follicle. Beneath the cortex lies the inner ovarian medulla, the site of blood vessels, lymph vessels and the nerves of the ovary.

The Ovarian Cycle

The formation of gametes in females is called oogenesis. The process begins with the ovarian stem cells, or oogonia, which are formed during foetal development. Unlike spermatogonia in males, however, oogonia form primary oocytes in the foetal ovary before birth. The number of primary oocytes present in the ovaries declines from one to two million in an infant, to approximately 400,000 at puberty, to zero by the end of the menopause. These primary oocytes remain in a resting state until puberty, held inside a follicle in the ovary. Ovulation, the release of an oocyte from the ovary, marks the transition from puberty into reproductive maturity for women. From then on, until the menopause signals the end of a woman's reproductive years, ovulation occurs approximately once every 28 days.

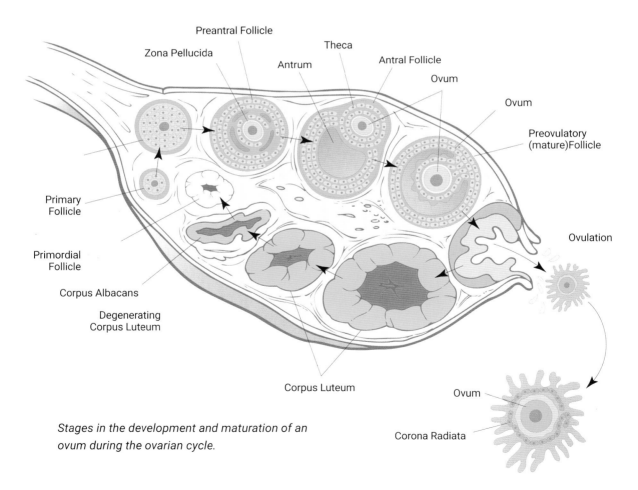

Stages in the development and maturation of an ovum during the ovarian cycle.

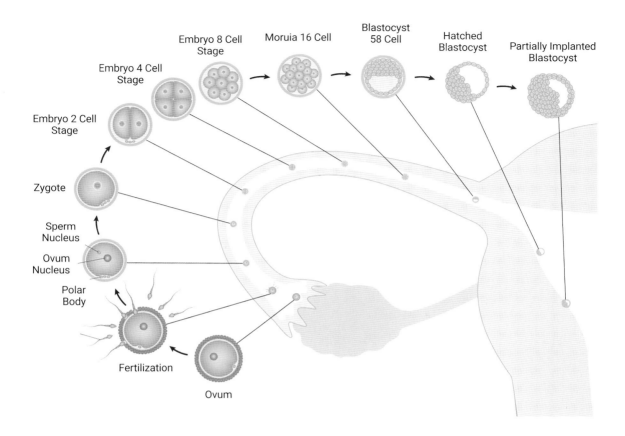

Embryo 8 Cell
Stage

Moruia 16 Cell

Blastocyst
58 Cell

Hatched
Blastocyst

Partially Implanted
Blastocyst

Embryo 4 Cell
Stage

Embryo 2 Cell
Stage

Zygote

Sperm
Nucleus

Ovum
Nucleus

Polar
Body

Fertilization

Ovum

The zygote formed after fertilization travels along the fallopian tube, growing and dividing until it becomes embedded in the wall of the uterus where the gestation of the foetus begins.

The first phase of the ovarian cycle, known as the follicular phase, generally lasts about 12 to 14 days for a 28-day menstrual cycle. During this phase, several ovarian follicles are stimulated to begin maturing, but usually only one, called the Graafian follicle, matures completely. The others will stop growing and disintegrate. Follicular development occurs because of a rise in the blood level of follicle-stimulating hormone (FSH), secreted by the pituitary gland. The maturing follicle releases oestrogen, the level of which rises throughout the follicular phase.

The second phase of the ovarian cycle is ovulation, which usually occurs around day 14. Just prior to ovulation, a surge of luteinizing hormone from the pituitary gland triggers the transition from primary to secondary oocyte. The primary oocyte divides, but not into two identical cells; one daughter cell is much larger than the other. This larger cell, the secondary oocyte, eventually leaves the ovary during ovulation. The smaller cell eventually disintegrates. Even though oogenesis may produce up to four cells, only one survives. The Graafian follicle bursts open and the secondary oocyte is released from the ovary to be swept into the nearby oviduct by waving, hair-like fimbriae.

The final phase of the cycle, which lasts about 14 days, is called the luteal phase. At this point the empty follicle develops into a structure called a corpus luteum. The corpus luteum produces large amounts of progesterone, a hormone which suppresses the production of FSH and luteinizing hormone and stimulates the continuing build-up of the endometrium. If pregnancy does not occur within 10 to 12 days, the corpus luteum stops secreting progesterone and eventually disintegrates in the ovary. If pregnancy does not take place the falling levels of progesterone cause the endometrium to break down and the next cycle begins with it being shed during menstruation. The redundant endometrium flows through an opening in the cervix, and out through the external opening of the vagina during menstruation.

Around the time of ovulation, the cervix undergoes changes that help sperm reach the ovum to fertilize it. The cervical canal widens, and cervical mucus becomes thinner and more alkaline. These changes help promote the passage of sperm from the vagina into the uterus and make the environment more hospitable to sperm. Fertilization of an ovum by a sperm normally occurs in an oviduct. In order for fertilization to occur, sperm must travel from the vagina where they are deposited, through the cervical canal to the uterus, and then on to one of the oviducts. Once sperm enter an oviduct, fluids carry them towards the secondary oocyte which releases molecules that attract the sperm and allow the

THE PLACENTA

The placenta is unique among the organs of the human body in that it only has a temporary existence. It is formed from cells in the blastocyst at the start of the pregnancy and therefore has the same genetic characteristics as the foetus. The placenta implants in the uterus and grows gradually during the first months of pregnancy; once completed it forms a spongy disc about 20 cm (8 in) in diameter and 3 cm (1.2 in) thick.

The function of the placenta is to supply nutrition and oxygen to the developing foetus and to remove wastes. It doesn't simply rest along the wall of the uterus. Tiny hair-like villi extend from the placenta into the uterine tissue holding it in place and opening up blood vessels in the uterine wall, facilitating blood flow from the mother to the placenta. Nutrients and oxygen cross from the maternal blood into the placental villi while wastes cross in the opposite direction. At the same time, the placenta creates a separation between the circulation of the mother and foetus, known as the placental barrier. The blood of the mother and the foetus never mix.

The foetus is connected to the placenta via the umbilical cord. Development of the umbilical cord begins around week 3 of pregnancy with the formation of the connecting stalk. The umbilical cord has fully formed by week 7, composed of the connecting stalk and umbilical vessels surrounding the amniotic membrane. The umbilical vein carries oxygenated blood with nutrients from the placenta to the foetus and the two umbilical arteries transport deoxygenated blood and waste products from the foetus to the placenta. On average, a developed umbilical cord is 50 to 60 cm (20–24 in) in length and 2 cm (0.75 in) in diameter.

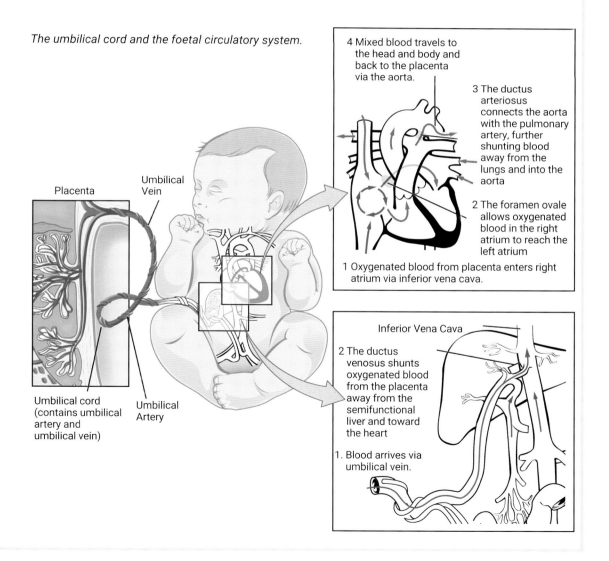

The umbilical cord and the foetal circulatory system.

Placenta

Umbilical Vein

Umbilical cord (contains umbilical artery and umbilical vein)

Umbilical Artery

4 Mixed blood travels to the head and body and back to the placenta via the aorta.

3 The ductus arteriosus connects the aorta with the pulmonary artery, further shunting blood away from the lungs and into the aorta

2 The foramen ovale allows oxygenated blood in the right atrium to reach the left atrium

1 Oxygenated blood from placenta enters right atrium via inferior vena cava.

Inferior Vena Cava

2 The ductus venosus shunts oxygenated blood from the placenta away from the semifunctional liver and toward the heart

1. Blood arrives via umbilical vein.

surface of the ovum to attach to the surface of the sperm. The sperm then enters the ovum, allowing fertilization to occur.

If the secondary oocyte is fertilized it divides, forming a diploid zygote and another polar body which normally breaks down and disappears. The zygote then continues the journey through the oviduct to the uterus, during which it undergoes several further divisions, so by the time it reaches the uterus around five days after fertilization, it has formed a ball of cells called a blastocyst. Within another day or two, the blastocyst implants itself in the endometrium lining the uterus, and gestation begins.

PREGNANCY AND CHILDBIRTH

The period of carrying one or more offspring by a woman from fertilization to childbirth is called pregnancy. The role of the mother is critical in the development of the offspring. She provides all the nutrients needed for normal growth and development of the offspring, and must also dispose of its wastes. Counting from the time of fertilization, pregnancy lasts 38 weeks, or about 40 weeks counting from the first day of the last menstrual period, though a normal pregnancy can last between 37 and 42 weeks.

The total duration of pregnancy is typically divided into three periods, called trimesters, each of about three months' duration. The first trimester begins at the time of fertilization. A woman in the first trimester is likely to experience signs and symptoms of pregnancy, such as a missed menstrual period, tender breasts, increased appetite, and more frequent urination. Many women also experience the nausea and vomiting of 'morning sickness'.

During the second trimester between weeks 13 to 28 the woman frequently feels more energized

THE MENTRUAL CYCLE

The menstrual cycle is necessary for the production of ova and the preparation of the uterus for pregnancy. It involves changes in both the ovaries and the uterus and is controlled by pituitary and ovarian hormones. Day 1 of the cycle is the first day of the menstrual period, when bleeding from the uterus begins as the thickened endometrium lining the uterus is shed. The endometrium builds up again during the remainder of the cycle and, if pregnancy does not occur, is shed again at the beginning of the next cycle. The development of a follicle in an ovary, ovulation of a secondary oocyte, and the subsequent degeneration of the follicle if pregnancy does not occur are all part of the menstrual cycle.

The length of the menstrual cycle may vary considerably. The average length of time between the first day of one menstrual period and the first day of the next period is 28 days, but it may range from 21 days to as many as 45 days. The menstrual period itself is usually about five days long, but it may vary from about two days to seven days.

The female reproductive years are defined by the commencement and cessation of the menstrual cycle. The first menstrual period, an event known as the menarche, usually occurs around 12 or 13 years of age, although there is considerable variation. It may occasionally occur as early as eight years or as late as 16. The menopause marks the cessation of menstrual

Opposite: Events in the female reproductive cycle.

cycles at the end of a woman's reproductive years. The average age of menopause is 52 years, but it may happen at any age between about 45 and 55 years of age.

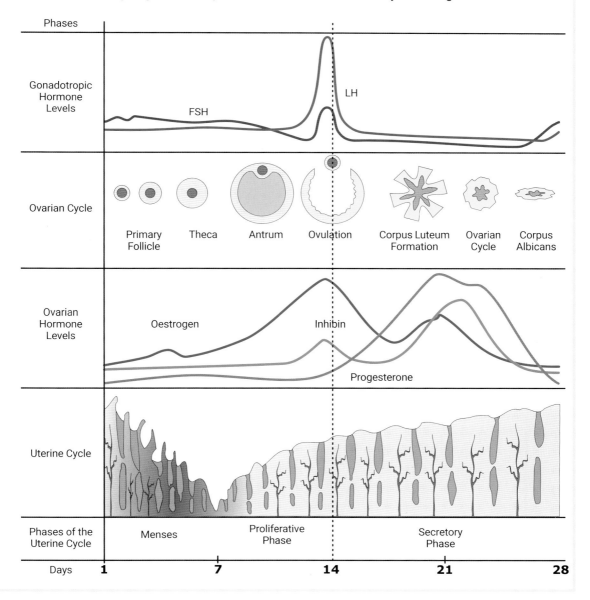

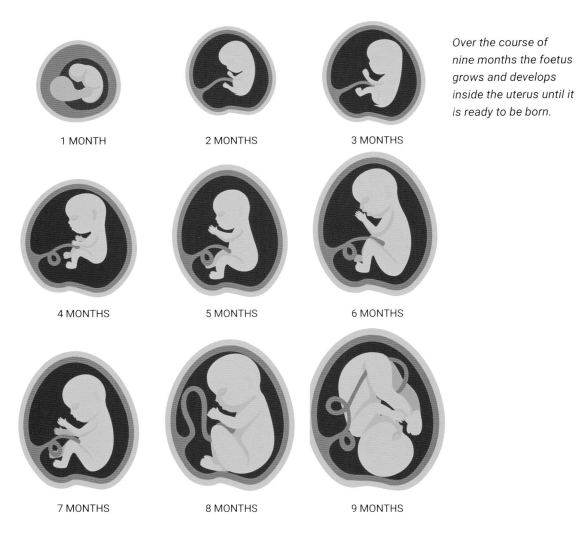

1 MONTH

2 MONTHS

3 MONTHS

Over the course of nine months the foetus grows and develops inside the uterus until it is ready to be born.

4 MONTHS

5 MONTHS

6 MONTHS

7 MONTHS

8 MONTHS

9 MONTHS

and starts to gain weight. By about week 20, the foetus is sufficiently developed that the mother can feel its movements. The third trimester from weeks 29 until birth at about 40 weeks sees weight gain increase and the uterus expand rapidly. The movements of the foetus become stronger and more frequent, and its growing weight and size may lead to symptoms such as back pain, swelling of the lower extremities, more frequent urination, varicose veins, and heartburn in the mother. By the end of the third trimester, the foetus turns to a downward position before birth, so its head rests on the cervix. This reduces the woman's bladder capacity and increases pressure on the pelvic floor and rectum.

Near the time of birth, the amniotic sac, the fluid-filled membrane enclosing the foetus within the uterus, breaks, releasing a gush of amniotic fluid – the event commonly described as 'the waters breaking'. Labour is the term for the process of childbirth in which powerful uterine contractions push the foetus and the placenta out of the body. It usually begins within a day of the waters breaking although it may begin prior to it. Labour can be divided into three stages: dilation, birth and afterbirth.

During the dilation stage, uterine contractions begin, becoming increasingly frequent and intense. The contractions push the baby's head against the cervix, causing the cervical canal to dilate to about 10 cm (4 in) in width, a process which may take 12 to 20 hours, or even longer.

The cervical canal m ust be dilated to this extent in order for the baby's head to fit through it.

The birth stage sees the baby descend (usually head first) through the cervical canal and vagina, and into the world outside. This stage may last from about 20 minutes to two hours or more. Usually, within a minute or less of birth, the umbilical cord is cut, so the baby is no longer connected to the placenta. During the afterbirth stage, the placenta is delivered. This stage may last from a few minutes to half an hour.

If there are problems that mean a normal vaginal delivery is not possible, for example the baby is in the breech (feet first) position and can't be turned, a caesarean section, or c-section may be carried out. This is a surgical procedure in which the baby is delivered through an incision made in the mother's abdominal wall and the wall of the uterus.

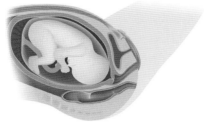

Foetal Expulsion

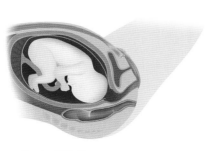

Cervical Dilation

Delivery Of Placenta

The three stages of birth – dilation of the cervix, birth of the baby and delivery of the placenta. The baby usually descends head first, though breech deliveries are not uncommon.

Glossary

ALLERGY: a response by the body's immune system to a substance such as plant pollen or a foodstuff that is generally harmless.

ALVEOLUS: (plural alveoli) tiny, thin-walled sac in the lungs where oxygen diffuses from the lungs into the bloodstream and carbon dioxide diffuses from the bloodstream into the lungs.

ANATOMY: the structure of the body and the study of that structure.

ANTIBODY: a type of protein produced by white blood cells in response to antigens on foreign substances stimulating the immune system.

ANTIGEN: identifying markers, usually proteins, on the surface of a cell or virus.

AORTA: the main artery of the circulatory system.

APPENDICULAR SKELETON: part of the skeleton that includes the bones of the limbs, shoulders and pelvis.

APPOCRINE GLAND: a sweat gland found in the armpits and groin that empties into a hair follicle.

ARTERIOLE: a blood vessel intermediate in size between and linking an artery and a capillary.

ARTERY: a blood vessel that carries blood away from the heart.

ATP: adenosine triphosphate, a molecule that stores and provides the energy needed to power cellular functions.

AUTONOMIC NERVOUS SYSTEM: Nerves connecting the central nervous system to the muscles and glands of the internal organs.

AXIAL SKELETON: part of the skeleton that includes the bones of the skull, spinal column, ribs and sternum.

B CELL: a type of lymphocyte that produces antibodies.

BLOOD-BRAIN BARRIER: protects the brain and spinal cord by controlling which substances enter the cerebrospinal fluid.

BLOOD TYPE: a characteristic determined by antigens on the surface of red blood cells; these differ between individuals and must be taken into consideration in the case of blood transfusions to avoid triggering an immune response.

BOWMAN'S CAPSULE: the part of a nephron that collects water and solutes filtered from the blood.

BRONCHIOLE: a finely branched airway inside the lung.

BURSA: a thin sac filled with lubricating fluid found near joints.

CAPILLARY: a blood vessel with a small diameter and thin walls through which oxygen, carbon dioxide and dissolved nutrients are exchanged between the bloodstream and the cells of the body.

CARDIAC CYCLE: the sequence of muscle contraction and relaxation in a single heartbeat.

CARDIOMYOCYTE: one of the muscle cells of the heart.

CARPALS: the eight bones of the wrist.

CELL: the smallest structural and functional unit of life.

CENTRAL NERVOUS SYSTEM: part of the nervous system that includes the brain and spinal cord.

CEREBELLUM: the hindbrain region responsible for coordinating limb movements.

CEREBRAL CORTEX: surface layer of the brain where sensory information is received, analysed and stored and responses coordinated.

CEREBROSPINAL FLUID: clear fluid that bathes and protects the brain and spinal cord.

CEREBRUM: the largest part of the brain, subdivided into lobes with various functions.

CHROMOSOME: a thread-like structure found in the nucleus of a cell that carries genetic information in the form of genes.

COLLAGEN: a type of protein that is a major component of tissues such as tendons and skin.

COMPLEMENT SYSTEM: a collection of proteins found in the blood that play a role in the immune system.

CONE CELL: a sensory cell in the retina of the eye that contributes to sharp daytime vision and colour perception.

CONNECTIVE TISSUE: tissue that connects and supports other tissues and organs in the body.

CORPUS LUTEUM: a hormone-secreting structure that forms from a ruptured ovarian follicle.

CYTOKINE: a protein secreted by white blood and other cells as part of the immune response.

DENDRITE: short, slender extension from the cell body of a neuron that collects incoming impulses.

DERMIS: layer of the skin beneath the epidermis.

DISSECTION: the act of cutting apart a body to study how it works.

DNA: deoxyribonucleic acid; the molecule of inheritance that encodes instructions for synthesizing proteins.

ECCRINE GLAND: a sweat gland that empties through pores on to the skin's surface.

ENDOCRINE GLAND: gland that secretes hormones into the bloodstream.

ENDOMETRIUM: the inner lining of the uterus.

ENDOSTEUM: a layer of cells around the central cavity of a bone responsible for growth and repair of the bone.

ENZYME: a type of protein that acts as a catalyst to regulate the speed of chemical reactions in cells.

EPIDERMIS: the surface layer of the skin.

EPIGLOTTIS: a flap of tissue between the pharynx and larynx that directs air into the trachea and food into the oesophagus.

EPIPHYSEAL PLATE: the site in the long bones of the body where bone lengthening takes place.

EPITHELIAL TISSUE (ALSO EPITHELIUM): tissue that forms body surfaces and lines cavities and organs throughout the body.

ERYTHROCYTE: a red blood cell.

EXOCRINE GLAND: a gland that secretes a product through ducts or tubes on to an epithelial surface.

FIBROBLAST: a type of cell that makes collagen and other constituents of connective tissue.

FOLLICLE: small sac or cavity such as that around the base of a hair; an oocyte and its surrounding layer of cells within an ovary.

GLOMERULUS: part of a nephron where water and solutes are filtered from the blood.

GLUCAGON: hormone secreted by the pancreas that stimulates the conversion of stored glycogen into glucose.

GLUCOSE: the simplest form of sugar, used by the cells of the body as a source of energy.

GLYCOGEN: the form in which glucose is stored in the body, usually in the liver and muscles.

GONAD: main reproductive organ in which sex cells, sperm or eggs, are produced.

GREY MATTER: cell bodies, dendrites and axons of neurons in the central nervous system that lack myelin sheaths.

HAEMOGLOBIN: an iron-containing protein in red blood cells that combines with oxygen to transport it around the body.

HAPLOID: a cell, generally a sperm or egg cell, that has only a single set of chromosomes rather than the two sets usually present.

HAVERSIAN CANAL: a microscopic tube that carries blood vessels within bone tissue.

HELPER T CELL: a type of lymphocyte that acts to activate other cells in the immune system to take action against pathogens.

HAEMOSTASIS: process that prevents blood loss from damaged blood vessels involving coagulation of blood and the formation of platelet plugs to seal the damage.

HINDBRAIN: part of the brain including the cerebellum that controls basic functions such as blood circulation and breathing and also helps to coordinate motor responses.

HOMEOSTASIS: the processes by which a stable internal environment is maintained within the body.

HORMONE: a chemical produced by an endocrine gland that is carried in the bloodstream to trigger a response in a target organ elsewhere in the body.

HYPODERMIS: innermost layer of the skin that provides insulation and an anchoring point for the upper skin layers.

HYPOTHALAMUS: region of the forebrain that controls processes such as body temperature, salt balance and blood pressure and triggers responses such as hunger and thirst.

IMMUNITY: resistance to a disease or pathogen.

INSULIN: a hormone that regulates glucose levels in the blood. Type 1 diabetes results from a lack of insulin; type 2 diabetes results from an inability to properly utilize insulin.

INTERSTITIAL FLUID: the fluid that occupies the space between cells in the body.

KERATINOCYTE: a cell that produces keratin, a tough structural protein; the most common type of cell in the epidermis of the skin.

LANGERHANS CELL: a macrophage cell found in the epidermis of the skin that is part of the immune system.

LARYNX: tubular airway leading to the lungs that contains the vocal cords.

LIGAMENT: a strap of dense connective tissue that bridges a joint or connects two bones.

LIMBIC SYSTEM: a system of centres in the brain that govern emotions and are involved in memory.

LOOP OF HENLE: hairpin-shaped tube in a nephron where water and solutes are reabsorbed.

LYMPH: tissue fluid that drains into the lymphatic system.

LYMPH NODE: part of the lymphatic system that plays a role in the immune response; lymphocytes inside the nodes clean up lymph before it reaches the bloodstream.

LYMPHOCYTE: a type of white blood cell that forms part of the immune system.

MACROPHAGE: a large white blood cell involved in the detection and destruction of bacteria and other invading organisms.

MEDULLA OBLONGATA: part of the hindbrain that coordinates basic tasks such as respiration; also plays a role in sleep and arousal.

MELANOCYTE: a cell in the epidermis of the skin that produces the pigment melanin.

MEMBRANE: a thin sheet of tissue covering a body surface.

MENOPAUSE: the end of a woman's reproductive years when her menstrual cycles cease.

METABOLISM: the sum of the biochemical processes essential to life that take place within the cells of the body.

MICROVILLI: finger-like projections on the surface of some cells that act to increase their surface area.

MITOCHONDRION (PLURAL MITOCHONDRIA): an organelle found in most cells where the processes of chemical respiration and energy production take place.

MOTOR NEURON: a neuron specialized to relay commands from the central nervous system to muscles and glands.

MYELIN SHEATH: an insulating covering around a sensory or motor neuron that enhances the transmission of nerve impulses.

NEGATIVE FEEDBACK: mechanism in homeostasis in which a condition that has changed triggers a response that reverses the change.

NEPHRON: the basic unit of the kidney which filters water and solutes from the blood and selectively reabsorbs them to maintain a homeostatic balance.

NEURON: a nerve cell, the basic unit of the nervous system.

NEUROGLIA: a class of cells that form part of the nervous system, providing a support structure for the neurons.

NEUROTRANSMITTER: a signalling molecule that passes an impulse from one neuron to another.

NEUTROPHIL: the most abundant and fastest

acting of the white blood cells; it attacks bacteria.

OOCYTE: an immature egg.

OOGENESIS: process by which a germ cell develops into an oocyte.

OSSICLE: one of the three small bones found in the inner ear.

OSTEOBLAST: a type of cell involved in the formation of new bone tissue.

OSTEOCLAST: a bone tissue cell that breaks down and reabsorbs old bone tissue.

OSTEOCYTE: an osteoblast that has completed its bone-forming function; the most common type of bone cell.

OSTEON: the basic unit of bone tissue, formed by a ring of osteocytes embedded around a central canal.

OVIDUCT: one of a pair of ducts through which eggs travel from an ovary to the uterus; also known as a fallopian tube.

OVUM: a mature oocyte.

PARASYMPATHETIC NERVOUS SYSTEM: part of the autonomic nervous system that tends to slow bodily activities and divert energy to basic tasks. Acts in opposition to the sympathetic nervous system.

PATHOGEN: any virus, bacterium or other organism that can infect the body and cause disease.

PERIOSTEUM: a layer of connective tissue covering a bone.

PERIPHERAL NERVOUS SYSTEM: all nerves leading into and out of the central nervous system.

PERISTALSIS: waves of muscular contraction and relaxation that push something through a tube such as the digestive tract.

PHAGOCYTE: a white blood cell that captures and envelops an invading bacterium or other organism.

PHALANGES: the 14 bones of the fingers.

PITUITARY GLAND: a gland of the endocrine system sometimes called the 'master' gland because it controls the activities of many other endocrine glands.

PLACENTA: organ only found during pregnancy that allows exchange of nutrients and waste between mother and foetus without direct contact between their bloodstreams.

PLASMA: the liquid part of blood comprised mainly of water with sugars, proteins, and other dissolved substances.

PLASMA MEMBRANE: a layer of lipids and proteins that forms the external boundary of a cell.

RESPIRATION: the exchange of oxygen from the air with waste carbon dioxide from cells through the process of breathing.

ROD CELL: receptor in the retina that is sensitive to dim light.

SEBACEOUS GLAND: a skin gland that produces sebum, a waterproof coating for hair and skin.

SENSORY RECEPTOR: specialized cell that can detect a stimulus.

SINEW: a tendon or ligament; a tough band of

connective tissue that links muscle to bone.

SOMATIC NERVOUS SYSTEM: nerves that lead from the central nervous system to the skeletal muscles, controlling their movement.

SPERMATOGENESIS: the formation of mature sperm from germ cells.

SPHINCTER: a circular band of muscle that opens and closes a passageway in the body.

SUTURE: a narrow immovable joint between the bones of the skull.

SYMPATHETIC NERVOUS SYSTEM: part of the autonomic nervous system that deals with increasing bodily activities at times of heightened danger or awareness; works in opposition to the parasympathetic nervous system.

SYNOVIAL FLUID: a clear fluid that acts to lubricate joints in the body.

SYNOVIAL JOINT: the most common type of joint in the body, enclosed within a cavity lubricated by synovial fluid.

T CELL: a type of lymphocyte that plays an important role in the immune system.

TARSALS: the seven bones of the ankle.

TENDON: a flexible strap of dense connective tissue attaching a muscle to bone.

THALAMUS: a region of the forebrain responsible for coordinating sensory inputs.

THYMUS GLAND: an organ of the lymphatic system where lymphocytes multiply and mature.

THYROID GLAND: endocrine gland located in front of the trachea that affects growth and development.

TRACHEA: windpipe through which air passes between the larynx and bronchi.

URETHRA: tube that conducts urine out of the body from the urinary bladder.

URINE: a fluid formed in the kidneys consisting of excess water and dissolved wastes.

UTERUS: a muscular, pear-shaped organ found in females in which an embryo is nurtured and grows during pregnancy.

VASOCONSTRICTION: the narrowing of a blood vessel by contracting small muscles in its walls.

VASODILATION: the widening of a blood vessel by relaxing the small muscles in its walls.

VEIN: a blood vessel that carries blood back to the heart.

VENULE: a blood vessel intermediate in size between and linking a vein and a capillary.

VILLUS (PLURAL VILLI): a finger-like projection from the surface of an epithelium, such as the lining of the small intestine, that increases its surface area.

WHITE MATTER: part of the brain and spinal cord where the nerve axons have myelin sheaths and specialize in rapid signal transmission.

ZYGOMATIC ARCH: the cheekbone.

Index

PICTURE CREDITS

t = top, b = bottom, l = left, r = right

Alamy: 7, 127, 129

Science Photo Library: 100, 109, 125, 131, 201

Shutterstock: 6, 8, 9, 15, 18, 20, 21, 22, 23, 26, 34, 36, 38, 40, 41, 44, 46, 48 (x2), 49, 51, 54, 56, 59, 65, 67, 69, 76, 77, 79, 84, 85, 86, 87, 88, 89, 90, 91, 94, 98, 99, 101, 102, 103, 104, 106, 108, 110, 112, 114, 115, 120, 121, 122, 123, 124, 126 (x2), 132, 137, 141, 142 (x2), 143 (x2), 145, 146, 147, 148, 151, 153, 154, 156, 161, 164, 165, 166, 170, 171, 172, 175, 178, 179, 180, 181, 183, 184, 187, 190, 194, 199, 200, 202, 203, 205, 206, 209, 213, 214, 215, 217, 219, 220, 221, 222, 227, 228, 230, 231, 233, 234, 235, 236, 238, 239, 243, 244, 245

Wellcome Collection: 11, 13, 14l

Wikimedia Commons: 10, 12, 14r, 24 (x2), 29, 47, 57, 60, 61, 62, 63, 64, 71, 72, 75, 82, 83, 97, 107, 128, 133, 135, 144, 150, 162, 173, 174, 182, 186, 188, 195, 196, 198, 216, 226, 241